Contemporary Trends and Issues in Science Education

Volume 52

The book series *Contemporary Trends and Issues in Science Education* provides a forum for innovative trends and issues connected to science education. Scholarship that focuses on advancing new visions, understanding, and is at the forefront of the field is found in this series. Accordingly, authoritative works based on empirical research and writings from disciplines external to science education, including historical, philosophical, psychological and sociological traditions, are represented here.

More information about this series at http://www.springer.com/series/6512

Maria Evagorou • Jan Alexis Nielsen
Justin Dillon
Editors

Science Teacher Education for Responsible Citizenship

Towards a Pedagogy for Relevance through Socioscientific Issues

Springer

Editors
Maria Evagorou
Department of Education
University of Nicosia
Nicosia, Cyprus

Jan Alexis Nielsen
Department of Science Education
University of Copenhagen
Copenhagen, Denmark

Justin Dillon
University of Exeter
Exeter, UK

ISSN 1878-0482 ISSN 1878-0784 (electronic)
Contemporary Trends and Issues in Science Education
ISBN 978-3-030-40228-0 ISBN 978-3-030-40229-7 (eBook)
https://doi.org/10.1007/978-3-030-40229-7

This Springer imprint is published by the registered company Springer Nature Switzerland AG.
The registered company address is: Gewerbestrasse 11, 6330 Cham, Switzerland

Foreword

I can remember it quite distinctly – the sound of a succession of steel gates each being locked behind me before the next one in sequence became unlocked for me to pass through. Needless to say, I was well aware of the high-powered rifles that scanned the courtyard as well as the stares and catcalls from inmates pressing weights as I made my way to the "school," a structure which served as the education building deep inside this maximum-security prison. The guards addressed the convicts as "Inmate 20065" or "Convict 14680." I addressed each by their name. There were Jack, Jesús, Robert, and Gerardo, among 46 others. I addressed each of them as though they were my students. In fact, they *were* my students.

It was 1978, and I found myself teaching a course for Syracuse University on Bioethics at Auburn Correctional Facility in upstate New York. Syracuse offered courses through a program called the independent study degree program, but the courses for the prison were actually offered face-to-face. This was one of the two assistantships I had to pay my way through graduate school. It was certainly a first for several reasons: the first time that I had ever set foot inside a prison, the first time that my students would be taking a college-level course, and the first time that a course in Bioethics had ever been offered at Syracuse – either on the main campus or in any other alternative setting.

Sounds quaint, doesn't it? An alternative setting. This was indeed an *alternative* setting. The charges ranged from murder and armed robbery to possession of narcotics and breaking and entering. It was also my first attempt at developing a responsible pedagogy, contextualized in science, aimed at developing citizenship. For such crimes, one could argue that a punitive, retributive, or reformative approach to incarceration was each appropriate. Perhaps some combination of those might bring about some measure of justice. My personal choice, both as a citizen and as an instructor, was to focus on the latter, which is one of the reasons I accepted this position. If and when any of my students would be released from prison, would they become members of a pluralistic society who would hopefully function as "productive" and "responsible" members of the human condition? Would they be reformed? Educated?

I saw my challenge as one in which my pedagogical decisions would need to be embedded in a sense of responsible pedagogy if my distal aim was the promotion of responsible citizenship. This was my opportunity to delve into areas of normative ethics, meta-ethics, moral reasoning, moral development, character development, a just community approach to education, argumentation, and related areas in the context of science, which were to become the precursors of the socio-scientific issues (SSI) framework some years later.

As I write this, in July 2019, it is difficult to fathom that 40 years has passed. I have just finished reading a brilliant collection of chapters from Maria Evagorou, Jan Alexis Nielsen, and Justin Dillon, who edited this book, *Science Teacher Education for Responsible Citizenship: Towards a Pedagogy for Relevance Through Socio-scientific Issues.* They have worked with some 37 international scholars to contribute chapters to this volume for the series "Contemporary Trends and Issues in Science Education." It would be difficult to find a stronger and more qualified group of contributors in the field of science teacher education.

The power and utility of this work lie in the global perspectives brought to bear on the topic of SSI as it relates to science teacher education. The overarching aims of utilizing the SSI framework are to cultivate responsible citizenship, to stress the importance of making science relevant to students, and, I might add, to compel science teacher educators to exercise pedagogically just decisions in promoting a functional view of scientific literacy that requires the exercise of evidence-based reasoning with a virtue ethic of care and compassion about the quality of our world and those that dwell within it. Accordingly, the contributors of this volume expand on issues related to promoting responsible citizenship, epistemological beliefs about teaching SSI, how SSI can play a central role in science teacher education and professional development, and challenges and the means to overcome them in designing and enacting SSI-based curricula. I found the chapters to be synergistically connected, each fleshing out the other, providing perspectives from young and seasoned scholars and all with an appropriate blend of empirical or analytical scholarship coupled with practical suggestions and implications for science teacher education and classroom practice. I do offer several leading questions that may serve as an advanced organizer when reading this book:

1. Do science teacher educators have a moral imperative to promote citizenship education?
2. How might science teacher educators prioritize and highlight the co-generative relationships among socio-scientific issues, subject matter knowledge, nature of science, and citizenship concerns for social and environmental justice?
3. What practices in science teacher education, and in the teaching of SSI, facilitate or potentially impede its pedagogical impact?
4. How can SSI best be taught as a method of guided inquiry?
5. How can SSI teaching help to dissipate the perceived barriers between the quadrivium and trivium?

For those pondering a possible dissertation or are so inclined to conduct further research, two addition questions are:

6. How can you build on this body of work to better articulate its goals in practice?
7. What are the conceptual or empirical blind spots in this body of work?

The irony was lost on me in 1978. It was not until a few years later that I learned the etymology of the word "school." It is derived from the Latin word *schola* as well as from the Greek word *skhole*. Both imply a sense of leisure and idleness (in the sense that one is not bound to have to work). In ancient times, if one was not compelled to work constantly in order to survive, then that person was certainly a member of the leisure class – a person of privilege who could spend their free time learning the seven liberal arts (i.e., quadrivium and trivium), including mathematics, music, rhetoric, and the like. One could participate in discussion and argumentation and participate in leading oneself out of ignorance. The irony, of course, was that those serving time had "time" as it were on their hands. Those who were selected to enroll in college coursework, to be schooled as it were, had leisure time to move from the house of brutes into one of higher consciousness. They could be led from ignorance and become educated, which is, after all, the root of the word "education." (To this day, I take a certain measure of delight in letting my university students know that they *not* working while they attend classes as that time is, in fact, their leisure time.)

So, I view science teacher education as a means to bring about responsible teachers who, in turn, bring about responsible students seeking to become virtuous in their words and deeds. This is nothing short of an educative exercise in participatory citizenship, and this is precisely what SSI and those who advocate SSI in science teacher education, such as the authors in this book, strive to achieve.

Distinguished University Professor
University of South Florida,
Tampa, FL, USA

Dana L. Zeidler

Contents

Contributors

Ruth Amos University College London, Institute of Education, London, UK

Alberto Bellocchi Queensland University of Technology, Brisbane, Australia

John Lawrence Bencze OISE, University of Toronto, Toronto, ON, Canada

Kennedy K. H. Chan Faculty of Education, The University of Hong Kong, Hong Kong, Hong Kong

Rachel Cohen Department of Science Teaching, Weizmann Institute of Science, Rehovot, Israel

Digna Couso CRECIM, Barcelona, Spain
Department Didàctica de les Matemàtiques i les Ciències Experimentals, Universitat Autònoma de Barcelona, Barcelona, Spain

James P. Davis Queensland University of Technology, Brisbane, Australia

Justin Dillon Graduate School of Education, Exeter University, Exeter, UK

Sarah El Halwany OISE, University of Toronto, Toronto, ON, Canada

Anna Garrido Espeja CRECIM, Barcelona, Spain

Maria Evagorou Department of Education, University of Nicosia, Nicosia, Cyprus

Patricia J. Friedrichsen Department of Learning, Teaching, and Curriculum, University of Missouri, Columbia, MO, USA

Melina Furman Universidad de San Andrés, Victoria, Argentina
CONICET, Buenos Aires, Argentina

Marie-Christine Knippels Freudenthal Institute, Utrecht University, Utrecht, Netherlands

Jessica S. C. Leung Faculty of Education, The University of Hong Kong, Hong Kong, Hong Kong

Ralph Levinson University College London, Institute of Education, London, UK

Mariana Luzuriaga Universidad de San Andrés, Victoria, Argentina

Ronicka Mudaly School of Education, University of KwaZulu – Natal, Durban, South Africa

Jan Alexis Nielsen Department of Science Education, University of Copenhagen, Copenhagen, Denmark

María Eugenia Podestá University of San Andrés, Victoria, Argentina

Troy D. Sadler University of North Carolina at Chapel Hill, Chapel Hill, NC, USA

Inés Taylor Universidad de San Andrés, Victoria, Argentina

Ka Lok Wong Faculty of Education, The University of Hong Kong, Hong Kong, Hong Kong

Anat Yarden Department of Science Teaching, Weizmann Institute of Science, Rehovot, Israel

Eran Zafrani Department of Science Teaching, Weizmann Institute of Science, Rehovot, Israel

Laura Zangori University of Missouri, Columbia, MO, USA

Majd Zouda OISE, University of Toronto, Toronto, ON, Canada

List of Figures

List of Tables

Chapter 1
Introduction: Socio-scientific Issues as Promoting Responsible Citizenship and the Relevance of Science

Maria Evagorou and Justin Dillon

1.1 Introduction

Socioscientic issues (SSI) and teacher professional development have been part of our research agendas for at least the last decade, and as we (Maria, Jan and Justin) crossed our paths as researchers, we started working together on a European Commission funded project titled *Preparing Science Educators for Everyday Science* (PreSEES). The aim of the project was to prepare pre-service science teachers (PST) in their effort to teach SSI, and with a group of researchers from around Europe we shared our ideas, questions and concerns and designed modules aiming to introduce SSI; help PSTs design and teach SSI related activities; and support PSTs in assessing learning in SSI (Evagorou et al. 2014a, b, also Chap. 10 in this book). During the project we realized that the international science education research community mostly explored how students engage in SSI (Patronis et al. 1999; Sadler et al. 2007; Sadler and Zeidler 2004; Sampson et al. 2011; Simonneaux and Simonneaux 2008; Shoulders and Myer 2013), but studies on teachers, their practices and how they can support their students to engage in SSI was still limited (Evagorou and Puig 2017; Tidemand and Nielsen 2016). At the same time in Europe, where all three of us live and work, various reform documents (EU Commission 2015; Owen et al. 2012) were shifting the emphasis of research agendas to responsible citizenship and the notion of making science relevant to students. The need to make science relevant to students came from reports showing that the numbers of students choosing science as a future career, or being interested in science, was

M. Evagorou (✉)
Department of Education, University of Nicosia, Nicosia, Cyprus
e-mail: evagorou.m@unic.ac.cy

J. Dillon
Graduate School of Education, Exeter University, Exeter, UK
e-mail: J.S.Dillon@exeter.ac.uk

M. Evagorou et al. (eds.), *Science Teacher Education for Responsible Citizenship*, Contemporary Trends and Issues in Science Education 52,
https://doi.org/10.1007/978-3-030-40229-7_1

declining (EC 2004), whilst the shift to responsible citizenship is linked to the idea of scientists and society sharing the outcomes of science in an effort to promote responsible research and innovation (Owen et al. 2012). Based on our understanding of SSI literature, we support that by engaging students and teachers in SSI we can actually achieve making science relevant, and promote responsible citizenship, and therefore as part of our on-going discussions and our findings from the PreSEES project we decided to explore the issue and invite researchers working on SSI and teacher professional development to present their work as part of this edited volume.

The purpose of this book is to bring together researchers working on teacher professional development, with an emphasis on SSI, to share their work, experiences and findings in terms of the pedagogies and pedagogical designs, and teaching strategies they are applying in their work in order to promote relevance of science and responsible citizenship. Therefore, our invitation was extended to researchers around the world, making an effort to include work from all continents to represent the abundance of work, and possibly different approaches in the different educational systems. The call for chapters was targeted to researchers whose work we were familiar with and believe represent different pedagogical and methodological approaches in introducing SSI to teachers. In this chapter we briefly present SSI and its importance, the importance of preparing science educators to teach SSI, and introduce the various chapters that are included in this book.

1.2 Socioscientific Issues, Relevance of Science and Responsible Citizenship

Socioscientific (SSI) issues are ill-structured problems that involve moral, ethical, and financial aspects, and lack clear-cut solutions (Lee and Grace 2012; Topcu et al. 2010), are usually topics that emerge from the nexus of science and society (Sadler and Zeidler 2004), and have a degree of uncertainty. Some of the topics include stem cell research, environmental issues and their possible solutions (i.e. fracking, renewable energy) and genetically modified organisms. According to Zeidler et al. (2005), by discussing such topics the SSI movement aims to empower students to consider how decisions made concerning science-based issues reflect "the moral principles and qualities of virtue that encompass their own lives, as well and the physical and social world around them" (p. 360). The ability to deal with socioscientific issues has been recognized as an important goal of science education (Zeidler 2014) and by engaging learners with SSI, we can potentially help them understand the relevance of science to their lives (Stuckey et al. 2013), understand aspects of the nature of science and how people use it, and develop their capacity to be critical consumers of scientific information (Kolstø 2001; Levinson 2006). Hence, SSI education is either concerned with ethics and involves moral judgment about issues of scientific concern, or represents those social issues and problems that are conceptually influenced by science and require the integration of science concepts and processes (Sadler et al. 2007). According to Simonneaux and Simonneaux (2008) when we teach SSI we aim:

> to improve knowledge understanding, to contribute to citizenship education, to help students to make an informed decision, to empower them to participate in debates, to help them to be able to deal with complexity, and to understand better the nature of science (p. 181).

Therefore, socioscientific issues enable learners to recognize that there is a human dimension to the practice of science, see the connections of science to everyday life (Zeidler et al. 2003) and take action on issues relevant to their everyday lives (Bencze and Carter 2011). The effort to introduce socioscientific issues in the curriculum was first evident as part of the Science Technology Society (STS) movement as far as back in the 1980s by making school science more reflective of the society (Sadler 2004). However, the main difference between STS and SSI, is that SSI focuses on empowering students to handle the socioscientific issues by understanding the various aspects of the issue, and making informed decisions (Kolstø 2001). Introducing SSI in science teaching can also be supported by Roberts' (2007) Vision II of science which aims to promote the understanding of the usefulness of scientific knowledge by using meaningful content (more about Vision II can be found in Chaps. 3, 4 and 9). More recently, European research agendas have been placing an emphasis on Responsible Research and Innovation (RRI), which is "an approach that anticipates and assesses potential implications and societal expectations with regard to research and innovation, with the aim to foster the design of inclusive and sustainable research and innovation" (EU Commission 2015). The main emphases of RRI is on science with and for society (Owen et al. 2012), and highlights the importance of presenting the societal aspects of science to our learners (Evagorou and Puig 2017) and inviting them to take action. With the two movements (RRI and SSI) sharing the common goal of discussing and understanding the societal aspects of science, researchers link RRI to socioscientific issues and responsible citizenship (Owen et al. 2012). The main ideas behind RRI (also presented in Chaps. 4 and 8), and socioscientific issues is that by including socioscientific issues in science learning and teaching we could move science classes towards unwrapping and engaging discussions about the intersections of science and society, promote scientific practices, and potentially invite students to act responsibly and participate actively. This "knowing in action" (Aikenhead 2006) aspect of RRI and SSI is related to what Sjöström and Eilks (2018) define as Vision III scientific literacy – one that includes socio-political action and moral-philosophical perspectives. Therefore, the inclusion of socioscientific issues in the curriculum offers a means of expanding both the curriculum and the range of instructional practices commonly experienced in the school science classroom, and in some cases also involves taking action on issues of concern.

Responsible citizenship, or thinking in scientifically responsible ways requires the students to form specific features of characters, and, "depends upon the character of both the scientists and the public at large, and that character includes reflexive judgment applied to scientific knowledge and ethical standards" (Zeidler et al. 2014,

p. 83). According to the same researchers, our character and experiences are shaped by our interactions in multiple cultures, and since science education is considered a distinct culture (Aikenhead 2006) we aim through socioscientific education to form characters that are ready to act in scientifically responsible ways (EU Commission 2015). Despite the effort to include SSI in schools, according to Alsop and Bencze (2012) SSI instruction is still constrained to a presentation of the social dilemma, with no attempt to promote engagement, participation or action. There is minimal research in the area of teacher education and the pedagogy associated with the relevance of science and responsible citizenship through their use (Evagorou and Puig 2017). Studies have shown that teachers do not make the connection between science and students everyday life since they find it difficult to relate scientific data and the social aspects of the problem which bring uncertainty into the discussions (Evagorou 2011; Forbes and Davis 2007). Research studies suggest that science teachers find it challenging to guide student learning in SSI (Evagorou 2011; Levinson 2006) and this is mainly because teaching SSI puts a demand on science teachers to use information and knowledge from outside their scientific domains (i.e., moral, financial, ethical dilemmas) (Simonneaux and Simonneaux 2008). Therefore, teacher professional development programs have started focusing on the pedagogical challenges of teaching SSI (Forbes and Davis 2007), but so far there is limited research on SSI teaching as part of either in-service or pre-service teacher education and this is the gap that we seek to address in this edited volume.

1.3 The Structure of the Book

This edited volume consists of 12 chapters. In Chaps. 1 and 2 we (as editors) introduce the notion of socioscientific issues, responsible citizenship and relevance of science and briefly discuss research on teacher professional development and SSI. In Chaps. 3, 4, 5, 6, 7, 8, 9, 10, and 11 we bring together the work of researchers from around the world (Australia, Argentina, Canada, Hong Kong, Israel, Spain, South Africa, UK, USA) to share their practices, examples of pedagogical approaches in teacher education (both pre- and in-service), and experiences on how to promote relevance and responsible citizenship through SSI. The final chapter concludes by summarizing new perspectives for addressing socioscientific issues in teacher education.

When considering the order of the chapters we decided against grouping them based on the type of professional development (pre- or in-service), or on the type of methodologies used. Instead, we decided to start with Chap. 2 as an overview of recent empirical research in SSI, follow with Leung and colleagues who set the basis by exploring teachers beliefs about teaching SSI and connecting them to Vision 2. The chapter to follow (Amos and colleagues) includes references to Vision 2 and then each subsequent chapter is linked to the following in a similar way.

In Chaps. 1 and 2 we (Maria, Jan and Justin) explain our views of socioscientific issues, and discuss recent research on teacher professional development and SSI. In our Chapters we use the term socioscientific instead of socio-scientific (see Zeidler 2014 for explanation) but we did not require the same from the Chapter authors.

In Chap. 3 (Pre-service secondary science teachers' beliefs about teaching socio-scientific issues), Leung, Wong and Chan examined 18 PSTs beliefs about the importance of SSI teaching in the local science curriculum and identifies their key learning experiences during a course on Nature of Science and SSI. Their results suggest that at the end of the course, most of the PSTs considered SSI as a key component of the science curriculum. In addition, the data analysis revealed three reasons why the PSTs did not prioritize SSI teaching in the curriculum: the complexity of SSI teaching, the shared curricular objectives of SSI with subjects (e.g. Liberal Arts in the case of Hong Kong) and their belief that SSI does not have an important role in content knowledge (CK) and nature of science (NOS). Furthermore, the results show that having a unidirectional view about the relationship between SSI, NOS and CK could lead PSTs to consider SSI teaching as less important to teaching CK and NOS. Therefore, Leung, Wong and Chan highlight the need to address the interrelationship between SSI, NOS and CK. An important aspect of this chapter is that the authors explain the context of Hong Kong and that even though the country adopted the STS movement in science teaching, SSI are not included in the science curricula, but are included instead in the Liberal Arts. Furthermore, an important finding that is highlighted in the discussion of the authors is that of the lack of assessment on SSI, which according to some PSTs is an important reason leading them to not include SSI in their teaching.

In Chap. 4 (Socio-scientific inquiry-based learning: possibilities and challenges for teacher education), Amos, Knippels and Levinson explore the implementation of a pedagogical approach for teaching through socially responsible inquiry embedded in socioscientific issues with pre-service teachers. More specifically, the work presented in this chapter comes from an European Commission funded project focusing on elaborating pedagogies which bring together, under the umbrella of Responsible Research and Innovation, the following: inquiry based science education (IBSE), earning of socio-scientific issues (SSI) and incorporating Citizenship Education (CE). More specifically, the authors of the chapter present how three pre-service teachers implemented socioscientific lessons. The studies showcase the complex steps involved in implementing SSI in the classroom, and more specifically the steps the pre-service teachers need to go through before they use science knowledge or structure inquiry based teaching. According to the authors, pre-service teachers need special scaffolding for these steps. An interesting aspect of this chapter is the teaching framework they present that is based on SSI and inquiry based learning.

In Chap. 5 (Critical and Active Public Engagement in Addressing Socioscientific Problems Through Science Teacher Education), Bencze, Halwany and Zouda present three collaborative case studies of science educators implementing the STEPWISE programme in different educational contexts. Their three very diverse case studies explain how different science educators, working with students of different ages and backgrounds are trying to incorporate Science Technology and Society and Nature of Science in their teaching. Important aspects of this chapter and the work by Bencze and colleagues is that: (a) they explain the differences between SSI and STS and (b) their work is more on the activist side with students taking actions for the problems they are exploring, making the connection in that way to responsible citizenship. Also, the way they work with science educators to support them is different from most of the other chapters, with long term collaborations and self-selected teachers.

In Chap. 6 (Supporting teachers in the design and enactment of socio-scientific issues based teaching in the US), Friedrichsen, Sadler and Zangori describe how using a collaborative professional development (PD) design supports teachers as they co-design and implement SSI curricula. In this chapter the authors present three case studies of how their work evolved over 3 years through the collaborative PD process. In the first case study they worked with an exemplar secondary school teacher to co-design curriculum materials placing an emphasis on SSI and also on the modeling practice. The second case study focused on working with a group of 19 secondary school teachers from diverse backgrounds (biology, chemistry, environmental science) and the third case study focused on working with elementary school teachers in a school to co-design and teach a unit on the monarch butterfly, using the modeling practice as part of the curriculum. An important aspect of this chapter is that the authors are collaborating with exemplar teachers and together are co-designing lessons. Another note is that the emphasis of this work is also on practices and the Next Generation Science Standards (Achieve 2013), with an emphasis also on the content. The authors discuss how including different practices when engaging students with SSI can work, or impede the teaching.

In Chap. 7 (Gamification of SSI's as a Science Pedagogy), Davis and Bellocchi propose a gamification approach as a strategy for teaching SSI with the aim to enhance science literacy and critical rationality. More specifically, through their work that has developed over the years, James and Alberto focus on strategies and pedagogical tools that might assist teachers to become producers of SSI games to be implemented in schools, and the possibilities that their strategies contribute to critical rationality. James and Alberto work with pre-service teachers in Australia. In their definition of SSI they mostly focus on Socially Acute Questions (ASQ) (Simonneaux 2014). When they refer to games, they mostly refer to Alternative Reality Games which combine virtual with real situations. They describe the way in which they used the games with pre-service teachers, a way that promotes not only SSI but also the use of technology. The way the researchers introduce SSI to their pre-service teachers means that the teachers need to get actively involved to understand the topic and solutions, and in this way they promote activism, responsible citizenship, and they make science relevant to students.

In Chap. 8 (Science teachers as proponents of socio-scientific inquiry-based learning: From professional development to classroom enactment) Cohen, Zafrani and Yarden discuss responsible and active citizenship through the implementation of the a specific pedagogical framework (also discussed in Chap. 4) with upper-secondary school teachers in Israel. By implementing the SSIBL framework, according to the authors the teachers must have the required content knowledge of and about SSI, the pedagogical knowledge needed for SSI inquiry, and attitudes needed to prepare students to make informed actions on SSI. The authors present two case studies, one with a teacher expert in SSI, and one with a teacher novice in SSI, and discuss the different ways in which these teachers implement the framework. The teachers were involved in professional that had a duration of 4 days and within this time they had to also implement in their classes an SSI lesson. Based on the findings, teachers with expertise in SSI find it easier to implement lessons framework, probably as SSI is more in accordance to their personal beliefs about teaching and learning. In their chapter the authors emphasize especially on active citizenship and explain Vision 1 and Vision 2 of science and how active citizenship is closer to vision 2 but this is not exercised in schools.

In Chap. 9 (Getting ready to work with socio-scientific issues in the classroom: a study with Argentine teachers) Furman, Taylor, Luzuriaga and Podesta present a professional development program implemented with in-service Argentinian teachers to support them in implementing SSI in their teaching. The authors of this chapter focus on following three teachers over 1 year as they implement SSI with their students, and reflect on struggles they face. The initial workshop had a duration of 2, 3 h twilight sessions. The teachers who decided to continue working with the researchers co-developed their lessons with the researchers. The final program has a duration of 20 h. The first teacher (kindergarten teacher) reports that she would have not been able to implement the SSI lesson without the help of the researchers, the second teacher (chemistry teacher) who was used at giving lectures mostly found it more difficult to implement SSI but according to her students she became a more hands-on teacher in her pedagogies. She reported however that SSI interfered with the content she wanted to teach. The third teacher, a biology teacher who used to be a researcher and did not have a pedagogical background was more open and more able to implement SSI in her class. Furthermore, the authors suggest that "the successful and sustained incorporation of SSI approaches depends on teachers' content knowledge and a more social understanding of science. Furman and colleagues explain the context in Argentina and how teachers in their country have a view of science as facts (Vision 1) and not as active participation (Vision 2). This is similar to what is reported in other chapters about teachers in other countries (Hong Kong, Israel, UK).

In Chap. 10 (Introducing SSI in primary pre-service teacher education: scientific practices to learn the big ideas of science) Garrido and Couso discuss how they have implemented a pre-service program aiming to prepare elementary school teachers to teach SSI, and how these teachers implement SSI activities in their classes. Specifically, Garrido and Couso designed and implemented a research-based training for pre-service primary school teachers. The aim was to help the PSTs to understand SSI, and enable them to teach them. In their study, linked to the PrESEES project) three pre-service teachers designed and implemented SSI lesson plans in primary schools, and reflected on the process. The results highlight that the SSI context supports the development of more innovative lesson plans in aspects as introducing the problem, including the scientific content and using formative assessment. The findings from this study also highlight difficulties implementing the lessons, similar to the ones reported in Chaps. 6 and 9.

In Chap. 11 (Re-thinking the integration of socioscientific issues in life sciences classrooms within the context of decolonizing the curriculum) Mudaly discusses the integration of indigenous knowledge into the curriculum in South Africa, explaining the rationale for introducing socioscientific issues within the South African context. Specifically she uses tenets from critical pedagogy to explore how novice teachers developed and taught science lessons which focussed on socioscientific issues within the context of decolonising the curriculum. Data collected include teachers' lesson plans and implementation and reflections on the process. According to Mudaly, teachers identified socioscientific issues based on their in-depth knowledge of the socio-cultural contexts of learners and their communities. This is the only chapter in the book making special reference to indigenous contexts, and novice teachers.

1.4 Summarizing

The collection of chapters showcase how researchers from around the globe apply different approaches for professional development, with most of the approaches being context related. For example in the US (Chap. 6) professional development is driven by the reform efforts and the Next Generation Science Standards. In Chaps 4 and 8 the professional development is driven by European Commission reports, and funding schemes placing an emphasis on responsible research and innovation and active citizenship in science. Similarly, in Chap. 3 the professional development is influenced by how SSI is implemented in the science curriculum (or not, given that it is not included as part of science, but as part of Liberal Arts), and also by the fact that SSI is not included as part of students' assessment. In Chap. 11 the professional development is arising from the need to decolonize the curriculum in South Africa and find ways to include indigenous knowledge as part of teaching science.

Most of the examples in this edited volume focus on long term professional development and working with self-selected and exemplar teachers. We have also included research studies emphasizing on pre-service and in-service professional development, with one of the studies making special reference to novice teachers. Despite the differences in the context, there are also many similarities across the chapters. For example the pedagogical framework adopted in Chaps. 4, 8 and 6 is similar in the sense that they focus on inquiry-based learning, or scientific practices, and include experimentation as part of the teaching process. Likewise, the professional development approaches in Chaps. 5 and 7 are similar in the sense that they focus on active engagement with the problem and the solution. Finally, an important theme across the chapters is that that different teachers apply the SSI pedagogies in different ways, and this might differ because of their personal characteristics (i.e. teaching philosophies, different nature of science views, content understanding). Findings from all chapters are critically discussed and summarized in Chap. 13 to refer to new perspectives for addressing SSI in teacher education.

References

Achieve. (2013). *The next generation science standards* (pp. 1–3). Retrieved from http://www.nextgenscience.org/

Aikenhead, G. S. (2006). *Science education for everyday life: evidence-based practice*. New York: Teachers College Press.

Alsop, S., & Bencze, L. (2012). In search of activist pedagogies in SMTE. *Canadian Journal of Science, Mathematics and Technology Education, 12*(4), 394–408. https://doi.org/10.1080/14926156.2012.732256.

Bencze, L., & Carter, L. (2011). Globalizing students acting for the common good. *Journal of Research in Science Teaching, 48*(6), 648–669.

European Commission. (2004). *Europe needs more scientists* (pp. 1–24). Luxembourg: Office for Official Publications of the European Communities.

European Commission. (2015). *Science education for responsible citizenship* (pp. 1–88). Luxembourg: Office for Official Publications of the European Communities.

Evagorou, M. (2011). Discussing a socioscientific issue in a primary school classroom: The case of using a technology-supported environment in formal and nonformal settings. In *Socio-scientific issues in the classroom* (Vol. 39, pp. 133–159). Dordrecht: Springer. https://doi.org/10.1007/978-94-007-1159-4_8.

Evagorou, M., Albe, V., Angelides, P., Couso, D., Chirlesan, G., Evans, R. H., . . . Nielsen, J. A. (2014a). Preparing preservice science teachers to teach socioscientific (SSI) argumentation. Science Teacher Education, 69, 39–47.

Evagorou, M., Guven, D., & Mugaloglu, E. (2014b). Preparing elementary and secondary pre-service teachers for everyday science. *Science Education International, 25*(1), 68–78.

Evagorou, M., & Puig, B. (2017). Engaging elementary school pre-service teachers in modeling a socioscientific issue as a way to help them appreciate the social aspects of science. *International Journal of Education in Mathematics, Science and Technology, 5*(2), 113–123.

Forbes, C. T., & Davis, E. A. (2007). Exploring preservice elementary teachers' critique and adaptation of science curriculum materials in respect to socioscientific issues. *Science & Education, 17*(8–9), 829–854. https://doi.org/10.1007/s11191-007-9080-z.

Kolstø, S. D. (2001). Scientific literacy for citizenship: Tools for dealing with the science dimension of controversial socioscientific issues. *Science Education, 85*(3), 291–310.

Lee, Y. C., & Grace, M. (2012). Students' reasoning and decision making about a socioscientific issue: A cross-context comparison. *Science Education, 96*(5), 787–807. https://doi.org/10.1002/sce.21021.

Levinson, R. (2006). Towards a theoretical framework for teaching controversial socio-scientific issues. *International Journal of Science Education, 28*(10), 1201–1224. https://doi.org/10.1080/09500690600560753.

Owen, R., Macnaghten, P., & Stilgoe, J. (2012). Responsible research and innovation: From science in society to science for society, with society. *Science and Public Policy, 39*(6), 751–760. https://doi.org/10.1093/scipol/scs093.

Patronis, T., Potari, D., & Spiliotopoulou, V. (1999). Students' argumentation in decision-making on a socio-scientific issue: Implications for teaching. *International Journal of Science Education, 21*(7), 745–754. https://doi.org/10.1080/095006999290408.

Roberts, D. A. (2007). Scientific literacy/science literacy. In S. K. Abell & N. G. Lederman (Eds.), *Handbook of research on science education* (pp. 729–780). Mahwah: Lawrence Erlbaum Associates.

Sadler, T. D. (2004). Moral sensitivity and its contribution to the resolution of socio-scientific issues. *Journal of Moral Education, 33*(3), 339–358. https://doi.org/10.1080/0305724042000733091.

Sadler, T. D., & Zeidler, D. L. (2004). Patterns of informal reasoning in the context of socioscientific decision making. *Journal of Research in Science Teaching, 42*(1), 112–138. https://doi.org/10.1002/tea.20042.

Sadler, T. D., Barab, S. A., & Scott, B. (2007). What do students gain by engaging in Socioscientific inquiry? *Research in Science Education, 37*(4), 371–391. https://doi.org/10.1007/s11165-006-9030-9.

Sampson, V., Simon, S., Amos, R., & Evagorou, M. (2011). Metalogue: Engaging students in scientific and socio-scientific argumentation. In *Socio-scientific issues in the classroom* (Vol. 39, pp. 193–199). Dordrecht: Springer. https://doi.org/10.1007/978-94-007-1159-4_11.

Shoulders, C., & Myers, B. (2013). Socioscientific issues-based instruction: An investigation of agriscience students' content knowledge based on student variables. *Journal of Agricultural Education, 54*(3), 140–156. https://doi.org/10.5032/jae.2013.03140.

Simonneaux, L. (2014). From promoting the techno-sciences to activism – A variety of objectives involved in the teaching of SSIs. In L. Bencze & S. Alsop (Eds.), *Activist science and technology education* (pp. 99–111). Dordrecht, The Netherlands: Springer.

Simonneaux, L., & Simonneaux, J. (2008). Socio-scientific reasoning influenced by identities. *Cultural Studies of Science Education, 4*(3), 705–711. https://doi.org/10.1007/s11422-008-9145-6.

Sjöström, J., & Eilks, I. (2018). Reconsidering different visions of scientific literacy and science education based on the concept of Bildung. In Y. Dori, Z. Mevarech, & D. Baker (Eds.), *Cognition, metacognition, and culture in STEM education. Innovations in science education and technology* (Vol. 24). Cham: Springer.

Stuckey, M., Hofstein, A., Mamlok-Naaman, R., & Eilks, I. (2013). The meaning of "relevance" in science education and its implications for the science curriculum. *Studies in Science Education, 49*(1), 1–34. https://doi.org/10.1080/03057267.2013.802463.

Tidemand, S., & Nielsen, J. A. (2016). The role of socioscientific issues in biology teaching: From the perspective of teachers. *International Journal of Science Education, 39*(1), 44–61. https://doi.org/10.1080/09500693.2016.1264644.

Topcu, M. S., Sadler, T. D., & Yilmaz Tuzun, O. (2010). Preservice science teachers' informal reasoning about socioscientific issues: The influence of issue context. *International Journal of Science Education, 32*(18), 2475–2495. https://doi.org/10.1080/09500690903524779.

Zeidler, D. L. (2014). Socioscientific issues as a curriculum emphasis. In N. G. Lederman & S. K. Abell (Eds.), *Handbook of research on Science Education* (pp. 697–726). New York/Oxfordshire: Routledge.

Zeidler, D. L., Osborne, J., Erduran, S., Simon, S., & Monk, M. (2003). The role of argument during discourse about socioscientific issues. In *The role of moral reasoning on socioscientific issues and discourse in science education* (pp. 97–116). Dordrecht: Springer. https://doi.org/10.1007/1-4020-4996-X_6.

Zeidler, D. L., Sadler, T. D., Simmons, M. L., & Howes, E. V. (2005). Beyond STS: A research-based framework for socioscientific issues education. *Science Education, 89*(3), 357–377. https://doi.org/10.1002/sce.20048.

Zeidler, D. L., Berkowitz, M. W., & Bennett, K. (2014). Thinking (scientifically) responsibly: The cultivation of character in a global science education community. In *Assessing schools for generation R (responsibility)* (Vol. 41, pp. 83–99). Dordrecht: Springer. https://doi.org/10.1007/978-94-007-2748-9_7.

Maria Evagorou is an Associate Professor at the University of Nicosia, Cyprus. Her scholarly activity focuses on exploring and enhancing young students'and pre-service teachers' argumentation skills within socioscientific issues. More specifically, the emphasis of her work is on students' and teachers' talk when they engage in the discussion of SSI and aims to explore ways in which teachers can be supported in their effort to include SSI in their teaching s. She has worked as a principal investigator and senior researcher on various EU and local research projects, served as a member on the JRST editorial board, and as a strand coordinator for ESERA, and has published widely.

Justin Dillon is professor of science and environmental education at the University of Exeter. After studying for a degree in chemistry, he trained as a science teacher at Chelsea College and taught in six inner London schools. Justin joined King's College London in 1989 and was promoted to professor in 2009. He is editor-in-chief of Studies in Science Education and co-edits the International Journal of Science Education. In 2007, Justin was elected President of the European Science Education Research Association. He has co-edited 18 books including Becoming a Teacher, Bad Education and the International Handbook of Research on Environmental Education. Justin has published around 100 papers in peer-reviewed journals and almost the same number of book chapters.

Chapter 2
Teachers and Socioscientific Issues – An Overview of Recent Empirical Research

Jan Alexis Nielsen

2.1 Introduction

Teachers play a central role in determining the uptake and quality of socioscientific (SSI)[1] teaching – i.e., science teaching in which students are engaged with socioscientific issues (Forbes and Davis 2008; Zeidler et al. 2005). From much of the 'early' research (pre-2010) on SSI, in general, and on SSI-teaching, in particular, we get the sense that science teachers face many challenges. For example, SSIs and their pedagogical potentials in science teaching seem somewhat alien to teachers (Lazarowitz and Bloch 2005; Lee et al. 2006); the lack of time to adapt to a fundamentally new practice and the lack of materials may be a key challenge to the uptake of SSI-teaching (Sadler et al. 2006); and, last but not least, SSI-teaching may require a pedagogical repertoire most often only held by non-science teachers (Simonneaux and Simonneaux 2009). To be sure, full-fledged SSI-teaching will inevitably involve guiding students' argumentation or decision-making processes (Nielsen 2009) while they weave together and weigh incommensurate factors – such as information coming not just from the natural sciences (Nielsen 2010).

This chapter explores the most recent empirical (2016–) research on pre- and in-service teachers' approaches to, or thinking about, teaching SSI, and presents some of the main trajectories in the findings. The reported research was identified through searches for the string OR ('socioscientific issues', 'socio-scientific issues', 'SSI', 'societal issues') in the Web of Science and ERIC databases. Only peer-reviewed journal articles or book chapters were included and only publications that

[1] I will use the standard abbreviations, 'SSI' for socioscientific issues and 'SSI-teaching' for socioscientific teaching'

J. A. Nielsen (✉)
Department of Science Education, University of Copenhagen, Copenhagen, Denmark
e-mail: janielsen@ind.ku.dk

M. Evagorou et al. (eds.), *Science Teacher Education for Responsible Citizenship*, Contemporary Trends and Issues in Science Education 52,
https://doi.org/10.1007/978-3-030-40229-7_2

reported on *empirical* investigations of (pre-service) teachers in relation to SSI. This ruled out publications that reports on students in relation to SSI and how teachers and researchers co-design SSI-units.

While the process was indeed done systematically, the aim of this chapter is not to be a systematic review of the literature on (pre-service) science teachers and SSI. Indeed, the chapter aims at providing a backdrop for the rest of this book. In recent publications – such as the ones by Tidemand and Nielsen (2017) and Evagorou and Puig (2017) – one can get a general overview of the pre-2016 literature on the topic (bear in mind that by far the most literature on SSI reports on investigations about students, and not teachers).

2.2 Pre-service Science Teacher' Relations to SSI

There is a growing body of research on SSI and pre-service science teachers (PSTs). Much of the literature focuses on the delivery and/or impact of special workshops or programmes on SSI teaching that are woven into an existing teacher education courses – often in the frame of a (research) project (see, for example, Evagorou and Puig 2017; Kilinc et al. 2017a, b). Of course, the fact that competence development in terms of SSI teaching is often an addendum to existing teacher education is not just a feature of the most recent research (see, for example, Evagorou et al. 2014b).

So there seems to be a clear sense among teacher educators that learning to teach SSI-based activities is not necessarily something that comes naturally. Indeed, Genel and Topcu (2016) recently went against the main research design trend and elected to study Turkish PSTs as they went into practice *without* furnishing them with special workshops or programmes in order to simply investigate how equipped 'normal' PSTs were for teaching SSI in the classroom. Unsurprisingly the conclusions of Genel and Topcu were not positive: 'Almost all aspects of SSI-based instruction we covered in the study confirmed that PSTs' understanding, and practices of SSI are not adequate' (2016, p. 116). My hypothesis is that similar studies in many other countries and cultures would generate similar findings.

As an aside, it is worth mentioning that a significant portion of the recent research on PST and SSI is situated in Turkey (Cebesoy and Oztekin 2016; Cinici 2016; Es et al. 2016; Genel and Topcu 2016; Kilinc et al. 2017a, b; Kutluca and Aydin 2017; Ozturk 2017; Ozturk and Yilmaz-Tuzun 2017; Ural Keles and Aydin 2017; Yapicioglu and Kaptan 2017). The fact that Turkish scholars are so prolific in this area can, in part, be traced back to a curriculum reform that focusses on SSI, but which has not been adequately substantiated on the level of teacher education (Genel and Topcu 2016). Many of these studies, not just the ones from Turkey, end with a plea for a more focused introduction to SSI during science teachers' education.

Now, it seems that SSI needs to be addressed at the level of teacher education. But how? Some recent research indicates that both shorter workshop sessions or training programmes about SSI for pre-service science teachers have positive immediate effects, but often limited effects on the PSTs ensuing practice. For example, when working with 20 PSTs, Evagorou and Puig (2017) found that while a programme on SSI teaching had an immediate impact on the degree to which the PSTs were able to identify societal aspects of science and science related issues, the PSTs did not manage to make the societal aspects operational in their ensuing teaching (see also Evagorou et al. 2014a). Similarly, Kilinc et al. (2017a) conducted a longitudinal in-depth single case study in order to identify how and why a PST can change her beliefs about dialogic teaching about SSI. They found that while training workshops that focus on dialogic argumentation in teaching can positively influence a pre-service teacher's valuing of dialogic teaching, the ensuing teaching experiences, to some effect, revered this positive change. In a study of PSTs' informal reasoning about a given SSI, Ozturk and Yilmaz-Tuzun (2017) found that while the participating PSTs seemed to have enough topic specific knowledge to understand the science aspects of SSI, the PSTs were not immediately able to mobilise this knowledge in their own informal reasoning and decision-making.

There seems to be a connection between PSTs' epistemological beliefs, on the one hand, and their own engagement with, or thinking about, SSI, on the other. For example, N. Ozturk and Yilmaz-Tuzun (2017) documented that PSTs' epistemological beliefs seemed to correlate with their ability to engage in informal reasoning concerning SSI – coarsely put, PSTs who believe that scientific knowledge is certain are less inclined to present counter arguments in their informal reasoning. Interestingly, when querying PSTs about their beliefs about the teacher's role in SSI teaching, Kilinc et al. (2017b), found a paradox: while most respondents identified teacher roles that afford dialogic SSI teaching, many of those held an *absolutist* epistemology of science knowledge – that is, that 'knowledge is certain and is given by authorities' (p. 197) – which the authors argue in practice would favour a monologic teaching practice. This study then indicates that while being positive when *talking* about SSI, other factors such as beliefs may influence practices in a way that undermines full-fledged SSI-teaching.

The general findings from these recent studies are less than uplifting. To my mind, the field of science education is still on the search for a viable way to place an emphasis on SSI-teaching in teacher education programmes in a way that really enables teachers to bring SSI into their future classrooms. In particular, it seems to be an open question whether the main factor behind these less-than-ideal findings is the (lack of) status of SSI-teaching in teacher education programmes – recall that in all the above cases, SSI-aspects were auxiliary and more or less added into an existing teacher education programme.

2.3 Science Teachers' Relations to, and Experiences with, SSI

A general theme in the research on in-service teachers' approaches to, or thinking about, SSI-teaching concerns teachers' assessment practices in relation to SSI-activities. For example, when Steffen and Hossle (2017) investigated German biology teachers' reflections on how to assess students' SSI decision-making abilities, they found that while the teachers embraced SSI decision-making as being very relevant, recognised that high quality formative assessment is central in learning processes and were clearly aware of potential student abilities pertinent to SSI decision-making, the teachers deferred from focusing on these abilities. Instead they focused on students' mastery of biological content. This finding resonates with those of Tidemand and Nielsen (2017) that while SSI is an explicit part of the Danish biology curriculum in upper secondary school, biology teachers cannot really be said to assess students' SSI abilities in class or in oral examinations. Similarly, Christenson et al. (2017) found that Swedish science teachers avoid assessing anything more than students' mastery of disciplinary content when assessing students in SSI-teaching. So, while SSI-aspects are woven into the curricula in countries such as Germany, Denmark and Sweden, the relevant assessment criteria are most often not integrated in classroom assessment practices. This, of course, is a major obstacle for the uptake of SSI-activities. As we know, assessment is a key determinant for what and how teachers focus their teaching on (Harlen 2007).

The apparent avoidance of SSI-related assessment criteria in classroom assessment may have several different but interrelated causes. The teachers in the abovementioned study by Steffen and Hossle (2017), referred to a 'lack of assessment criteria' (p. 47) and they placed greater value on the 'usual assessment criteria' (p. 48) – leading the authors to suggest that the position taken by the teachers may be due to the fact that SSI-teaching had not yet been integrated into the German school practice. Other studies, such as the one by Christenson et al. (2017) and the one by Tidemand and Nielsen (2017), indicate that science teachers may simply not have enough experience in focusing on the SSI-relevant assessment criteria. Future research might investigate whether teachers are poised to exclude SSI-related assessment criteria because they are unfamiliar with assessing along such criteria or because they simply feel that such criteria are only marginally important.

Christenson et al. (2017) seized on the notion that teachers from other subjects may be more experienced with certain elements of SSI-teaching by comparing how science teachers and Swedish language teachers approached issues concerning the assessment of students in SSI-teaching. Their analysis indicates that both science and Swedish language teachers focused on students' mastery of disciplinary content, but that the Swedish language teachers *also* 'included students' abilities to use content knowledge by selecting and citing references, the structure of the argumentation, and, in addition, language' (p. 1416). Neither the science teachers nor the Swedish language teachers included students' ability to engage in considerations about the foundation of ethical principles. Based on their findings, Christenson et al. (2017) argue that science teachers may benefit in the long term from interdisciplinary collaboration with language teachers – but they also stated that this may be too time-consuming and that everything hinges on whether science teachers are able to accept that the radius of their discipline expands to include SSI.

Pitiporntapin and Srisakuna (2017) conducted a study of the development of three Thai science teachers – with no SSI-teaching experience – over a semester in a TPD programme on SSI-teaching. While they found that the teachers positively developed their SSI-teaching, the study highlighted the fundamental need of local support structures for sustaining SSI-teaching. Indeed, Pitiporntapin and Srisakuna (2017) found that while the teachers wanted to change their practice they were – as relative novices (in terms of SSI-teaching) – challenged by the fact that there was no one with more experience on SSI-teaching among their colleagues.

Leden et al. (2017) followed a group of science teachers over 3 years in order to, among other things, investigate changes in teachers' identification of issues (hereunder SSIs) for teaching and opportunities and challenges related to the teaching of these issues. The main intervention in the study consisted of 12 group discussions that were distributed over the 3 years and that were thematically focused on aspects of the nature of science. Leden et al. (2017) found that over time the science teachers notably became better at identifying potential issues for teaching, that the issues they identified went from being dominantly imprecise and general to being dominantly aligned with SSI-teaching, and that the teachers found more and more opportunities and benefits related to teaching such issues – such as 'as increased engagement and the development of critical thinking and reflexivity' on the side of the students (p. 500). Thus the study of Leden et al. (2017) suggests that it may be a good investment to take time out to allow teachers to collaboratively negotiate the meaning of, and reflect on their own practical experiences with, a SSI as a new teaching approach.

Leden et al.'s (2017) findings resonate with the findings in the study of Tidemand and Nielsen (2017), that without specific training in aspects concerning SSI teaching, Danish upper secondary biology teachers tend to elicit very general and imprecise issues as examples of SSI. Indeed, in the study by Tidemand and Nielsen (2017) the teachers interviewed tended to see *all* biology teaching as SSI because all biology content is *potentially* relevant to something in society. In the study by Tidemand and Nielsen (2017), this narrative may be seen as legitimising a content-centred approach to SSI-teaching, the origins of which can be traced back to a curriculum that, on the one hand, includes SSI-laden aspects and, on the other, focuses heavily on the student being able to exhibit knowledge about core biological content.

More prominent were Sund's (2016) findings from interviewing 29 science teachers from Swedish upper-secondary schools about what they value in and about science teaching. He found that 12 out of the 29 had closest affinity with a teaching tradition according to which 'the relation between facts and values is important and teachers offer situations in which students can develop abilities to use their knowledge in daily life and also at a societal level' (p. 401). Sund's (2016) refreshing study suggest that the outlook for the Swedish curriculum reform (which, in part, focuses on SSI) is less bleak than suggested by most other research studies internationally.

2.4 Conclusions

One of the key threads in the most recent research on pre- and inservice teachers' relation to SSI is the need for facilitating processes in which teachers (perhaps with guidance by trainers and/or researchers) work to make SSI-related learning objectives *operational* for assessment and teaching (Nielsen et al. 2018). The lack of a clearly operationalised notion of SSI is a specific episode of the general problem that generic or cross-curricular learning goals are often ill-defined and not sufficiently operationalised (Belova et al. 2017; Dolin et al. 2017; Nielsen and Dolin 2016).

Further, attempts to add SSI-elements onto existing teacher education curricula has had mixed results on teachers' views and competences. In so far as it is a political aim that future students develop their SSI decision-making abilities, it might be supported by a move to substantially integrate SSI aspects into teacher education programmes. But it is not yet clear what is needed in order to enable teachers to bring SSI into their future classrooms.

References

Belova, N., Dittmar, J., Hansson, L., Hofstein, A., Nielsen, J. A., Sjöström, J., & Eilks, I. (2017). Cross-curricular goals and raising the relevance of science education. In K. Hahl, K. Juuti, J. Lampiselkä, A. Uitto, & J. Lavonen (Eds.), *Cognitive and affective aspects in science education research: Selected papers from the ESERA 2015 conference* (pp. 297–307). Cham: Springer.

Cebesoy, U. B., & Oztekin, C. (2016). Relationships among Turkish pre-service science teachers' genetics literacy levels and their attitudes towards issues in genetics literacy. *Journal of Baltic Science Education, 15*(2), 159–172.

Christenson, N., Gericke, N., & Rundgren, S. N. C. (2017). Science and language teachers' assessment of upper secondary students' socioscientific argumentation. *International Journal of Science and Mathematics Education, 15*(8), 1403–1422. https://doi.org/10.1007/s10763-016-9746-6.

Cinici, A. (2016). Balancing the pros and cons of GMOs: Socio-scientific argumentation in pre-service teacher education. *International Journal of Science Education, 38*(11), 1841–1866. https://doi.org/10.1080/09500693.2016.1220033.

Dolin, J., Nielsen, J. A., & Tidemand, S. (2017). Evaluering af naturfaglige kompetencer. *Acta Didactica Norge, 11*(3), 1–28.

Es, H., Mercan, S. I., & Ayas, C. (2016). A new socio-scientific issue for Turkey: Life with nuclear. *Turkish Journal of Education, 5*(2), 47–59. https://doi.org/10.19128/turje.92919.

Evagorou, M., & Puig, B. (2017). Engaging elementary school pre-service teachers in Modeling a Socioscientific issue as a way to help them appreciate the social aspects of science. *International Journal of Education in Mathematics, Science and Technology, 5*(2), 113–123. https://doi.org/10.18404/ijemst.99074.

Evagorou, M., Albe, V., Angelides, P., Couso, D., Chirlesan, G., Evans, R. H., et al. (2014a). Preparing preservice science teachers to teach socio-scientific (SSI) argumentation. *Science Teacher Education, 69*, 39–47.

Evagorou, M., Guven, D., & Mugaloglu, E. (2014b). Preparing elementary and secondary preservice teachers for everyday science. *Science Education International, 25*(1), 68–78.

Forbes, C. T., & Davis, E. A. (2008). Exploring preservice elementary teachers' critique and adaptation of science curriculum materials in respect to socioscientific issues. *Science & Education, 17*(8–9), 829–854. https://doi.org/10.1007/s11191-007-9080-z.

Genel, A., & Topcu, M. S. (2016). Turkish preservice science teachers' socioscientific issues-based teaching practices in middle school science classrooms. *Research in Science & Technological Education, 34*(1), 105–123. https://doi.org/10.1080/02635143.2015.1124847.

Harlen, W. (2007). *Assessment of learning*. London: Sage Publications Ltd.

Kilinc, A., Demiral, U., & Kartal, T. (2017a). Resistance to dialogic discourse in SSI teaching: The effects of an argumentation-based workshop, teaching practicum, and induction on a preservice science teacher. *Journal of Research in Science Teaching, 54*(6), 764–789. https://doi.org/10.1002/tea.21385.

Kilinc, A., Kelly, T., Eroglu, B., Demiral, U., Kartal, T., Sonmez, A., & Demirbag, M. (2017b). Stickers to facts, imposers, democracy advocators, and committed impartialists: Preservice science teachers' beliefs about teacher's roles in socioscientific discourses. *International Journal of Science and Mathematics Education, 15*(2), 195–213. https://doi.org/10.1007/s10763-015-9682-x.

Kutluca, A. Y., & Aydin, A. (2017). Changes in pre-service science teachers' understandings after being involved in explicit nature of science and socioscientific argumentation processes. *Science & Education, 26*(6), 637–668. https://doi.org/10.1007/s11191-017-9919-x.

Lazarowitz, R., & Bloch, I. (2005). Awareness of societal issues among high school biology teachers teaching genetics. *Journal of Science Education and Technology, 14*(5–6), 437–457. https://doi.org/10.1007/s10956-005-0220-4.

Leden, L., Hansson, L., & Redfors, A. (2017). From black and white to shades of grey. A longitudinal study of teachers' perspectives on teaching sociocultural and subjective aspects of science. *Science & Education, 26*(5), 483–511. https://doi.org/10.1007/s11191-017-9920-4.

Lee, H., Abd-El-Khalick, F., & Choi, K. (2006). Korean science teachers' perceptions of the introduction of socio-scientific issues into the science curriculum. *Canadian Journal of Science, Mathematics and Technology Education, 6*(2), 97–117. https://doi.org/10.1080/14926150609556691.

Nielsen, J. A. (2009). Structuring students' critical discussions through processes of decision-making on socio-scientific controversies. *Revista de Estudos Universitários, 35*(2), 139–165.

Nielsen, J. A. (2010). Functional roles of science in socio-scientific discussions. In I. Eilks & B. Ralle (Eds.), *Contemporary science education – Implications from science education research about orientations, strategies and assessment* (pp. 83–96). Aachen: Shaker.

Nielsen, J. A., & Dolin, J. (2016). Evaluering mellem mestring og præstation. [Assessment between mastery and performance]. *MONA, 2016*(1), 51–62.

Nielsen, J. A., Tidemand, S., & Dolin, J. (2018). Transforming assessment research: Recommendations for future research. In J. Dolin & R. Evans (Eds.), *Transforming assessment: Through an interplay between practice, research and policy* (pp. 279–290). Cham: Springer.

Ozturk, A. (2017). An investigation of prospective science teachers' socio-scientific argumentation processes in terms of metacognition: A causal-comparative study. *Pegem Egitim Ve Ogretim Dergisi, 7*(4), 547–582. https://doi.org/10.14527/pegegog.2017.020.

Ozturk, N., & Yilmaz-Tuzun, O. (2017). Preservice science Teachers' epistemological beliefs and informal reasoning regarding socioscientific issues. *Research in Science Education, 47*(6), 1275–1304. https://doi.org/10.1007/s11165-016-9548-4.

Pitiporntapin, S., & Srisakuna, S. (2017). Case studies of the development of science teachers' practices of Socio-Scientific Issue (SSI)-based teaching through a professional development. *Program, 10*(1), 56–66.

Sadler, T. D., Amirshokoohi, A., Kazempour, M., & Allspaw, K. M. (2006). Socioscience and ethics in science classrooms: Teacher perspectives and strategies. *Journal of Research in Science Teaching, 43*(4), 353–376. https://doi.org/10.1002/tea.20142.

Simonneaux, L., & Simonneaux, J. (2009). Students' socio-scientific reasoning on controversies from the viewpoint of education for sustainable development. *Cultural Studies of Science Education, 4*(3), 657–687. https://doi.org/10.1007/s11422-008-9141-x.

Steffen, B., & Hossle, C. (2017). Assessing students' performances in decision-making: Coping strategies of biology teachers. *Journal of Biological Education, 51*(1), 44–51. https://doi.org/10.1080/00219266.2016.1156012.

Sund, P. (2016). Discerning selective traditions in science education: A qualitative study of teachers' responses to what is important in science teaching. *Cultural Studies of Science Education, 11*(2), 387–409.

Tidemand, S., & Nielsen, J. A. (2017). The role of socioscientific issues in biology teaching: From the perspective of teachers. *International Journal of Science Education, 39*(1), 44–61. https://doi.org/10.1080/09500693.2016.1264644.

Ural Keles, P., & Aydin, S. (2017). Class teacher candidates' opinions on Genetically Modified Organisms (GMO). *Journal of Education and Practice, 8*(12), 145–150.

Yapicioglu, A. E., & Kaptan, F. (2017). A mixed method research study on the effectiveness of Socioscientific issue-based instruction. *Egitim Ve Bilim-Education and Science, 42*(192), 113–137. https://doi.org/10.15390/eb.2017.6600.

Zeidler, D. L., Sadler, T. D., Simmons, M. L., & Howes, E. V. (2005). Beyond STS: A research-based framework for socioscientific issues education. *Science Education, 89*(3), 357–377. https://doi.org/10.1002/sce.20048.

Jan Alexis Nielsen is Associate Professor at the Department of Science Education at the University of Copenhagen, Denmark. His research focuses on the understanding and assessment of generic competences. He has investigated students' socioscientific argumentation and teachers' beliefs about teaching socioscientific issues. He has also worked closely with teachers to build assessment frameworks for inquiry, modelling and innovation competence. Over the past 12 years, he has participated in a range of international and national projects related to inquiry and/or socioscientific issues.

Chapter 3
Pre-service Secondary Science Teachers' Beliefs About Teaching Socio-scientific Issues

Jessica S. C. Leung, Ka Lok Wong, and Kennedy K. H. Chan

3.1 Background

Scientifically literate individuals should not only have rich scientific knowledge but also be able to make informed decisions and participate in public debates on scientific issues. As a result, over the last few decades, there has been a call to address socio-scientific issues (SSI) in science education (e.g., Sadler 2011; Zeidler et al. 2005). Despite this growing advocacy for SSI teaching in secondary education, there is a gap between this theoretical ideal and the current practice of teachers. Although Hong Kong followed the science-technology-society (STS) movement in the 1980s and 1990s and integrated science-technology-society-environment (STSE) into secondary science curricula (Curriculum Development Council & Hong Kong Examinations and Assessment Authority 2017), SSI remain absent in the science curriculum and public examination of science. In contrast, SSI have found their place in Liberal Studies (LS), a core subject for all senior secondary students in Hong Kong. LS includes one particular Area of Study called Science, Technology and the Environment, in which one of the objectives is to enable students to 'be aware of the social, cultural and moral issues related to science, technology and the environment […] and […] to make judgements and informed decisions on [the issues]' (CDC & HKEAA 2015, p. 43). In this context it is not surprising that SSI teaching is rare, if not absent, in Hong Kong science education. However, while the extent to which relevant scientific knowledge and evidence are properly discussed in LS classrooms remains questionable, science teachers can contribute to the desired aims of SSI teaching in Hong Kong.

This study addresses the discrepancy between what the reform promotes (i.e., SSI teaching in secondary science education) and the reality in Hong Kong (i.e., the

J. S. C. Leung (✉) · K. L. Wong · K. K. H. Chan
Faculty of Education, The University of Hong Kong, Hong Kong, Hong Kong
e-mail: leungscj@hku.hk; klwong3@hku.hk; kennedyckh@hku.hk

M. Evagorou et al. (eds.), *Science Teacher Education for Responsible Citizenship*, Contemporary Trends and Issues in Science Education 52,
https://doi.org/10.1007/978-3-030-40229-7_3

scarcity of SSI teaching in science classrooms) by focusing on pre-service science teachers. Our choice to focus on pre-service teachers (PSTs) was inspired by previous studies suggesting that less experienced teachers were more likely to use STS topics than their more experienced counterparts (Lumpe et al. 1998). Therefore, our study is situated in the context of a reform-oriented initial science teacher education course designed to improve pre-service science teachers' understanding of SSI. We believe that science educators should consider teachers' beliefs about the importance of teaching SSI to promote its implementation in science classrooms. As a result, the study examines PSTs' beliefs about the importance of SSI teaching in the local science curriculum and identifies their key learning experiences during the course. This information can help to improve not only our course design, but also that of other teacher education courses. The following research questions guided our study:

1. What were the PSTs' beliefs about the importance of SSI teaching in the science curriculum and the reasons for changing their beliefs, if any, after attending this course?
2. What did the PSTs find most impressive in this course to facilitate their learning about SSI teaching?

3.2 Theoretical Framework

In the following section, we first define SSI and emphasise their importance in the development of students' scientific literacy. We then present our rationale for focusing on PSTs' beliefs about SSI teaching in this study.

3.2.1 Socio-scientific Issues (SSI)

SSI refers to issues emerging from the interrelationship of science and society that are often factually and ethically complex, with no clear solution, subject to on-going inquiry and based on uncertain, fragile and conflicting evidence (Sadler 2004). An essential difference between SSI and STSE is that SSI pays attention to the ethical dimension of social issues with conceptual, methodological and/or technological links to science (Zeidler et al. 2005). Ratcliffe and Grace (2003, pp. 2–3) suggested that socio-scientific issues:

1. Have a basis in science, frequently that at the frontiers of scientific knowledge;
2. Involve forming opinions, making choices at personal or societal level;
3. Are frequently media-reported, with attendant issues of presentation based on the purposes of the communicator;
4. Deal with incomplete information because of conflicting/incomplete scientific evidence, and inevitably incomplete reporting;

5. Address local, national and global dimensions with attendant political and societal frameworks;
6. Involve some cost-benefit analysis in which risk interacts with values;
7. May involve consideration of sustainable development;
8. Involve values and ethical reasoning;
9. May require some understanding of probability and risk; and
10. Are frequently topical with a transient life.

Many studies have highlighted that teaching SSI can lead to desirable student learning outcomes, such as fostering critical thinking skills, decision making, argumentation, reflective judgement and moral development (e.g., Sadler 2004; Zeidler et al. 2011). More importantly, SSI teaching can help develop students' scientific literacy (Zeidler et al. 2005), which involves making informed decisions about SSI as a significant component (Zeidler 2014).

Roberts (2007) summarised various perspectives on scientific literacy and classified them into two main streams: *Vision I* and *Vision II*. *Vision I* explores science from an inward perspective, focusing on 'the products and processes of science itself' (p. 730). In contrast, *Vision II* adopts an outward approach by focusing on 'situations in which science can legitimately be seen to play a role in other human affairs' (p. 729) or 'character of situations with a scientific component, situations that students are likely to encounter as citizens' (p. 730). In other words, *Vision I* focuses on the understanding of science itself, while *Vision II* focuses on the role of science in human affairs, and there is a continuum between these two extremes.

Hodson (2003) suggested that '[t]raditionally, science education has dealt with established and secure knowledge, while contested knowledge, multiple solutions, controversy and ethics have been excluded' (p. 664). It is not surprising that many science teachers simply consider that their main task is to teach scientific principles and concepts, while any substantive pedagogical change is regarded as a burden (Lee and Witz 2009). However, given the complex, open-ended and value-laden nature of SSI and the involvement of values and ethical reasoning, addressing only *Vision I* in instruction is not enough to help students understand SSI.

3.2.2 Teachers' Beliefs About SSI

Teachers' beliefs play an important role in shaping teachers' practices in the classroom (e.g., Bryan and Atwater 2002; King et al. 2001). Lee and Witz (2009) indicated that teachers may choose to discuss SSI because of their beliefs and values. Several studies have investigated teachers' beliefs about the importance of teaching SSI. Kara (2012) examined the perceptions of 102 pre-service biology teachers in Turkey on SSI teaching using a questionnaire they designed. It was revealed that PSTs of biology generally believed that SSI should be taught in biology classrooms. In a similar study involving 37 science teachers in the US, Pedersen and Totten (2001) found that although teachers thought it was important to teach social issues in science classrooms, only some believed that teaching social issues was as important as teaching science content. In a more recent qualitative study, Tidemand and Nielsen (2017) examined Danish biology teachers' beliefs about the role and function of SSI teaching activities for biology teaching in upper secondary schools. The authors revealed that teachers tended to have a content-centred interpretation of SSI, as evidenced by their use of SSI as an instrument to engage and facilitate students' learning of content knowledge. Only a few teachers taught SSI to prepare students to deal with issues outside of school.

We believe it is essential for PSTs to develop their beliefs about the importance of teaching SSI. As a result, we examine their beliefs about the importance of SSI in the science curriculum and their underlying reasons, using a course focusing on SSI. We expect the position of SSI teaching in the science curriculum to reflect the importance of SSI in science education, in turn affecting *whether* and *how* SSI should be implemented in science classrooms.

3.2.3 Course Design

The course adopted a reflection orientation in its conceptualisation (Abell and Bryan 1997). Participants were frequently invited to reflect on their learning experiences during the course by writing a reflective journal. The course consisted of three modules over 12 weeks (24 contact hours). In terms of content, the course focused on improving PSTs' NOS conceptions including the epistemological and sociological aspects of science (McComas 1998), their understanding of SSI (Ratcliffe and Grace 2003) and the intertwining nature of NOS and SSI. The focus on both NOS and SSI was informed by previous studies suggesting that NOS provided students with relevant conceptual tools to critically examine specific SSI (e.g., Khishfe 2012; Leung et al. 2015) and that SSI served as an effective context to improve students' conceptions of NOS (Sadler et al. 2004). As a result, our design echoed Karisan and Zeidler's (2017, p. 148) suggestions that '[g]iven the corpus of research around NOS and SSI (Zeidler 2014), we also suggest that teacher training programs should be reformed to include the integration of NOS in the context of SSI'. In terms of pedagogical approach, the course intended to foster PSTs' learning through reflection from a learner perspective and a teacher perspective, as detailed below.

(1) **Reflection from a learner perspective.** We reproduced Demirdöğen, Hanuscin, Uzuntiryaki-Kondakci and Köseoğlu's (2016) pedagogical approach to teach NOS to improve PSTs' NOS conceptions and their instructional repertoires. Specifically, we adopted an explicit reflective approach (Abd-El-Khalick and Lederman 2000) and used activities, such as Post box activities (Hume 2009), to help students reflect on their new understanding of NOS. The instructors modelled how to run a debate session about nanofood, which engaged participants in socio-scientific reasoning (Sadler et al. 2017) and argumentation discourse.

(2) **Reflection from a teacher perspective.** We also explicitly addressed NOS and SSI from a teacher perspective. For instance, in Module 3, instructors discussed with students the principles and strategies of developing scientific literacy using news media in the classroom following Jarman and McClune (2007). Module 3 also included two instructional sessions with video analysis of authentic video footage featuring SSI and NOS teaching in Hong Kong classrooms (Yip et al. 2018). Table 3.1 summarises the structure, key ideas and activities adopted in the course.

3.3 Methods

The study used a qualitative case study approach (Merriam 2009). Eighteen PSTs (nine females and nine males (pseudonyms used below)) enrolled in the final year credit-bearing course entitled *Nature of Science and Socio-scientific Issues* voluntarily participated in the study.

3.3.1 Data Collection

Change in Teachers' Beliefs About SSI The following three questions were extracted from the written survey on Pedagogical Content Knowledge about Socio-scientific Issues (PCK-SSI) (Tosunoglu and Lederman 2016) administered immediately before the Module on *SSI as learners* in Week 5 to document participants' beliefs about SSI teaching:

1. Should SSI teaching be a part of the science curriculum?
2. Do you think that emphasizing SSI in the science curriculum is necessary? (If yes, why? And if no, why (not?)?)
3. Do you believe it is important to spend instructional time in your science classroom to teach students about SSI? (If yes, why? And if no, why not?)

Table 3.1 Summary of topics in each module

Weeks	Module	Key ideas	Pedagogical activities
1–4	1. NOS as learners	NOS: A philosophical, epistemological and socio-cultural perspectives	Post box activities
			Interactive discussion
5–8	2. SSI as learners	The nature of SSI; socio-scientific reasoning and synthesis of ideas and practices	Emergent graphical interpretation
			Interactive discussion
			Peer debate
9–12	3. NOS and SSI as teachers	The use of media for teaching; pedagogy and assessment of NOS and SSI	Interactive discussion
			Video analysis

At the end of the course, as part of their assessment, participants wrote an essay with the following instruction:

> In light of the latest developments in the local science curriculum, identify the components (e.g., science content knowledge, NOS, SSI) that you consider essential. Rank these components according to their level of importance and present your arguments in the form of a written essay.

The essays reflected their beliefs about the importance of the different components of the local science curriculum and the reasons for their beliefs. It is worth noting that participants did not necessarily include SSI as an essential component of the science curriculum. In any case, their reasons allowed us to explore their views and reasoning about SSI teaching in relation to other aspects of the science curriculum.

Key Learning Moments Participants kept a reflective journal to record their thoughts about SSI and its teaching and its change (if any) as well as their key learning moments after each module. The relevant parts of the journals (i.e., Modules 2 and 3) were used as the data source.

3.3.2 Data Analysis

The data collected were analysed qualitatively using open coding (Creswell 2008). The three authors reviewed the data corpus independently before meeting to discuss the initial codes and develop the consensus codes. The first author then re-read each student's responses and assigned codes to the data. The team re-examined the data with assigned codes until a consensus was reached. Validation strategies (Creswell 2007), such as investigator triangulation and frequent peer debriefing between the co-authors, were used to ensure the trustworthiness of the results. The following section describes in detail the analysis of the main constructs of the study.

Teaching Beliefs About SSI We analysed participants' responses to the PCK-SSI questionnaire, which was administered immediately before Module 2 to capture the influence of SSI-related Modules on PSTs' beliefs about SSI teaching (referred to as *pre-course* thereafter), and their written essays (post-course), which focused on their beliefs about the importance of SSI teaching in the science curriculum and their underlying reasons. As the analysis progressed, three categories of participant views, namely *instrumental view*, *beyond an instrumental view* and *others*, emerged. This classification was informed by the literature and an interaction with the data. The term *instrumental* was borrowed from Tidemand and Nielsen (2017) and described the use of SSI teaching activities as an *instrument* to motivate, frame or put into perspective the teaching of a given science content. The use of this term resulted from a content-centred interpretation of SSI teaching by teachers. As a result, *instrumental view* included the use of SSI to (1) motivate and stimulate students' interest in science learning (*motivation and interest*), (2) apply science content knowledge to real life (*knowledge application*), and (3) *facilitate the learning*

Table 3.2 Classification of teaching beliefs about SSI

Categories	Sub-categories	References
Instrumental view	Motivation and interest	Tidemand and Nielsen (2017)
	Knowledge application	Tidemand and Nielsen (2017)
	Facilitating learning of science content	Sadler et al. (2016)
Beyond an instrumental view	Citizenship education	Vesterinen et al. (2016)
	As a context for teaching NOS	Karisan and Zeidler (2017)
	Skill development (e.g., critical thinking skills, decision-making, argumentation, reflective judgement and moral development)	Sadler (2004) and Zeidler et al. (2011)
	Values education	Lee et al. 2013
	Development of scientific literacy	Zeidler et al. (2005)

of science content. Conversely, *beyond an instrumental view* referred to the use of SSI teaching activities for purposes other than acquiring content knowledge, e.g., *skill development*, *as a context for teaching NOS*, *citizenship education* and the *development of scientific literacy* (see Table 3.2).

Key Learning Moments The classroom activities perceived as useful by the participants for their learning and how these activities supported their learning, as illustrated by their reflective journals, were analysed qualitatively. The key learning moments relevant to SSI teaching were identified and categorised according to their primary focus, SSI from a learner perspective and SSI from a teacher perspective.

3.4 Results and Discussion

In this section, we first report the results of the pre-course and post-course participants' beliefs about the importance of SSI teaching in the science curriculum and their reasons for incorporating (or not) SSI into the curriculum (RQ1). We then provide an overview of the key learning moments identified by the participants and how these classroom activities contributed to their learning (RQ2).

3.4.1 Pre-course Teaching Beliefs About SSI

As shown in Table 3.3, all participants agreed that SSI teaching should be part of the science curriculum. Of the 18 participants, 15 agreed on the importance of emphasising SSI and 16 on spending instructional time on SSI teaching. Contrary to the

Table 3.3 Participants' beliefs about the importance of SSI in the science curriculum

	No. of participants		
	Agree	Disagree	Indecisive
SSI should be part of the science curriculum	18	0	0
SSI should be emphasised in the science curriculum	15	3	0
It is important to spend instructional time on SSI	16	1	1

majority view, three participants disagreed that the science curriculum should emphasise SSI and one did not consider it important to devote instructional time to SSI. The underlying reasons are discussed later.

Table 3.4 presents the reasons given by the participants for integrating SSI into the curriculum. Their views can be classified as follows: (1) *instrumental view*, (2) *beyond an instrumental view* and (3) *others*. For *instrumental view*, 6, 4 and 2 out of the 18 participants considered that SSI could offer opportunities to apply scientific knowledge, boost motivation and interest in science learning and facilitate science learning, respectively. In addition, 8 and 5 out of the 18 participants held a *beyond an instrumental view*, perceiving SSI as a context for learning NOS and developing skills. Some participants proposed reasons that could be classified as an *instrumental view* and *beyond an instrumental view* (e.g., Cheryl, Rick and Winnie), indicating that these two views were not mutually exclusive. Unlike Tidemand and Nielsen's (2017) results on in-service teachers which indicated the high prevalence of *instrumental view* among in-service teachers, only 6 out of the 18 participants gave reasons classified only as an *instrumental view*. This result may be attributed to the focus on SSI and NOS in this course. The two participants (Ian and Lillian) belonging to *others* proposed that Hong Kong should follow the global trend of SSI teaching and supported their beliefs with reasons classified as an *instrumental view* or *beyond an instrumental view*.

Due to the lack of explicit focus on SSI in the public examination, Joyce, one of the three participants, did not consider it necessary to focus on SSI. Her statement clearly illustrated her view:

> I think focussing on NOS and SSI in the science curriculum is not necessary, unless there is a corresponding assessment reflecting students' understanding of them… If the curriculum emphasises NOS and SSI but the hard work of teachers and students cannot be objectively reflected, this may mislead teachers when planning their lessons.

Rick shared a similar view with Joyce, albeit being more optimistic. He contended that SSI should be emphasised in the curriculum and the public examination so that teachers and students would be motivated to teach and learn about SSI. His view was reflected in the following statement:

> … emphasising NOS and SSI in the curriculum can encourage teaching and learning about these elements. Due to public examination rewards, teachers are more likely to incorporate NOS and SSI into their teaching, while students are more motivated to learn them.

Keith echoed Rick's suggestion that SSI should be emphasised in the science curriculum, but admitted that little instructional time could be spent on SSI due to their low importance in the public examination, as illustrated by his statement: 'I

Table 3.4 Pre-course reasons for integrating SSI teaching into the science curriculum

Reasons	No. of participants	Participants	Sample excerpts
Instrumental view			
Knowledge application	6	Cheryl[a], Ian[b], Keith, Lillian[c], Rick[a], Winnie[a]	*SSI is important to prepare students to apply scientific knowledge in society* (Keith).
Motivation and interest	4	Gladys, Ian[b], Sam, Wilson	*As SSI topics are interesting and relatable to students' lives, they can also encourage students to pursue scientific knowledge* (Sam).
Facilitating science learning	3	Daniel, Gladys, Winnie[a]	*SSI refers to controversial social issues related to science... Therefore, SSI provides a ground for an open-ended discussion to facilitate students' learning and understanding of science* (Winnie).
Sub-total	10		
Beyond an instrumental view			
As a context for NOS	8	Bianca, Charles, Cheryl[a], Ian[b], Joyce, Morris, William, Winnie[a]	*SSI is one of the tools with which students can apply their understanding of NOS and teachers can assess students' understanding or beliefs about NOS* (William).
Skill development	5	Anastasia, Rick[a], Tiffany, Wendy, Winnie[a]	*SSI encourages students to practise moral reasoning and critical thinking... The skills they acquire in science class, like critical thinking and reasoning, will also be applicable in the future* (Anastasia).
Sub-total	12		
Others			
Global trend	2	Ian[a], Lillian[b]	*NOS and SSI are part of the science curriculum of many countries. Therefore, they can be considered an essential part of science education* (Lillian).
Sub-total	2		

Note: [a]Participants holding an *instrumental view* and *beyond an instrumental view*; [b]participants holding an *instrumental view, beyond an instrumental view* and *others*; [c]participants holding an *instrumental view* and *others*

will definitely discuss SSI with my students, but as usually less than 5% of the public examination questions are about SSI, I will probably spend little time on it…'

His view was consistent with previous studies on in-service teachers suggesting that lack of instructional time for content with little or no coverage in examinations was often perceived as a barrier to SSI teaching (Lee et al. 2006). Given the exam-oriented culture in Hong Kong, participants indicated that they generally focused on preparing students for public examination, which put heavy weighting on content knowledge. In other words, their motivation to teach SSI depended largely on the curriculum and the weighting of SSI in the public examination. This result suggested that in the absence of curriculum and assessment reform, science teachers would have little or no incentive to teach SSI.

3.4.2 Post-course Teachers' Beliefs About SSI

At the end of the course, 17 out of the 18 participants considered SSI as one of the important components of the science curriculum (see Table 3.5). Of these 17 participants, 9 prioritised content knowledge (CK) over SSI in the science curriculum, placing them nearer to *Vision I* than *Vision II*. In contrast, the remaining 8 participants prioritised SSI over CK in the science curriculum, placing them nearer to *Vision II* than *Vision I*.

As shown in Table 3.6, 15 out of the 18 participants gave reasons justifying their prioritization of SSI teaching in the science curriculum. While only 4 proposed reasons indicative of an *instrumental view*, all these 15 participants proposed reasons classified as *beyond an instrumental view*, compared with 12 in the pre-course stage. In other words, none of them held a purely *instrumental view*, compared with 6 participants in the pre-course stage. These results suggested a shift from an *instrumental view* to one that went beyond it.

Table 3.5 Participants' ranking of the components of the science curriculum according to their level of importance

Ranking	1	2	3	4	5
Vision I-oriented					
Daniel	CK	NOS	SSI	–	
Morris	CK	NOS, SI and SL	–	–	SSI and STSE
Rick	CK	NOS	SSI	–	–
Sam	CK	NOS	SSI		
Wilson	CK	NOS	SSI	–	–
Keith	CK, NOS, STSE	–	–	SSI	–
Winnie	NOS	CK	SSI	–	–
Bianca	SL	CK	SI	NOS	SSI
Anastasia	STEM	NOS	CK	SSI	–
Vision II-oriented					
Charles	SSI	SI	NOS	CK	–
Cheryl	SSI and SI	–	NOS	CK	
Gladys	STSE	NOS, SSI	–	SI	CK
Lillian	STSE	NOS	SSI	–	–
William	SI	SSI	Unifying concepts	NOS	–
Tiffany	STEM	NOS, SSI	–	–	–
Wendy	STEM	NOS, SSI	–	CK	–
Ian	STEM	NOS	SSI	CK	–
Other					
Joyce	Scientific investigation	NOS	Information literacy	–	–

Note: *NOS* (nature of science), *SI* (scientific inquiry), *SL* (scientific literacy), *CK* (content knowledge), *SSI* (socio-scientific issues), *STEM* (science-technology-engineering-mathematics), STSE (science-technology-society-environment)

Based on their view of scientific literacy, participants were further categorised into three groups, namely *Vision I*, *Vision II* and *Others* (i.e., ranking neither SSI nor CK as an important component of the science curriculum). According to this categorisation, their proposed reasons for the importance of SSI in the curriculum were presented in an attempt to compare these reasons with their view of scientific literacy. Seven of the nine participants in the *Vision I* group and all eight participants in the *Vision II* group gave reasons to prioritise SSI teaching (see Table 3.6). For *beyond an instrumental view*, in addition to reasons related to skill development and as a context for NOS, three new supporting reasons were proposed, namely *citizenship education*, *values education* and the *development of scientific literacy*. It is worth noting that three quarters of *Vision II* participants prioritised SSI teaching for citizenship education, compared with just over a fifth of *Vision I* participants. This difference can be attributed to the alignment of citizenship education with their teaching beliefs.

Further examination of the different responses revealed some possible reasons why SSI should *not* be prioritised in the science curriculum. Consider Morris' response:

> Although SSI and STSE in the science curriculum can develop students' positive attitude towards the contribution of science to socio-scientific issues, covering SSI and STSE in detail in science class is challenging for teachers because of the complexity of the different issues.

The above excerpt highlighted the first reason – *the complexity of SSI and the associated challenges*. This view was consistent with previous studies on in-service teachers, which discussed the challenges faced by teachers in SSI teaching, including lack of knowledge about SSI, uncertainty about how to conduct controversial discussions and how to manage lessons using small-group discussions, role playing and similar teaching strategies (Bryce and Gray 2004; Lee et al. 2006). Science teachers in Hong Kong are used to teaching CK, which usually has absolute answers. Therefore, this result suggested that it would be a challenge for them to teach SSI, which is more complex, open-ended and value-laden.

Another reason specific to the Hong Kong context emerged, represented by the following excerpt from Keith:

> SSI may be less important due to the presence of LS, which is a core subject in the local curriculum […] LS teachers may further 'connect knowledge and concepts across different disciplines' (CDC & HKEAA 2015) and this may provide a more well-rounded training than the SSI approach…

Keith's statement clearly illustrated the second reason – some participants believed that it would be *more appropriate to teach SSI in other subjects* (e.g., LS in the Hong Kong curriculum context). This view echoed previous findings from studies involving in-service teachers (Tidemand and Nielsen 2017). Although the current literature has largely corroborated these two reasons, we found a unique reason, as evidenced by Rick's statement: '[t]he reason for SSI's low ranking is that negotiating SSI somehow depends on science CK and understanding of NOS'. Rick believed in the more fundamental role of CK and NOS in supporting SSI negotia-

Table 3.6 Post-course reasons for prioritising SSI teaching

Reasons for prioritising SSI teaching	Number of participants (n = 15)				Excerpts
	Vision I (n = 9)		*Vision II* (n = 8)		
Instrumental view					
Knowledge application	2	Anastasia, Sam	2	Cheryl, Gladys	*... the integration of SSI in the curriculum... allow[s] students to make good use of scientific knowledge relevant to society...* (Anastasia, *Vision I*)
Sub-total	2		2		
Total	4				
Beyond an instrumental view					
Citizenship education	2	Anastasia, Rick	6	Cheryl, Charles, Gladys, Ian, Lillian, Wendy	Holbrook (2008)[1] *argued that based on the introduction of conceptual science, student enhancement of scientific literacy needs to consider a societal frame and to embrace the socio-scientific situation that provides the relevance for promoting responsible citizenship* (Wendy, *Vision II*).
Skill development	4	Anastasia, Daniel, Sam, Wilson	3	Cheryl, Gladys, Ian	*SSI can help students... make decisions based on evidence and help them think critically and consider moral and ethical reasoning* (Bybee et al. 2009)[2] (Sam, *Vision I*).
As a context for NOS	3	Anastasia, Wilson, Winnie	3	Gladys, Tiffany, William	*... what makes SSI irreplaceable is how they incorporate multiple outcomes, for example, scientific literacy and the nature of science, also referred to as its 'unification power'* (Zeidler et al. 2005)[4] (Anastasia, *Vision I*).
Development of scientific literacy	4	Anastasia, Bianca, Daniel, Rick	1	Tiffany	*Negotiating SSI in a science classroom can provide valuable experience for students to critically evaluate the arguments of different stakeholders and ultimately determine their own position in a complex situation. This is consistent with Vision II in scientific literacy* (Roberts 2007),[3] *which is related to literacy in science-related situations* (Rick, *Vision I*).
Values education	2	Winnie, Wilson	0	–	*SSI... plays an important role in the formulation of students' personal values in science education, as it offers students an opportunity 'to develop their personal values'... and 'judgements'...* (Wilson, *Vision I*)
Subtotal	7		8		
Total	15				

tion. More importantly, his statement emphasised his limited view of a *unidirectional* relationship between CK and NOS to help students negotiate SSI without realising how SSI could help CK and NOS learning in an intertwining manner.

Daniel's statement also reflected a similar belief: '[a]s SSI can be perceived as a context, it can be incorporated into other components, and there is no pedagogical need to teach and learn about SSI in a separate, decontextualised way. Daniel suggested that SSI should be viewed as context rather than content. Nevertheless, unlike Rick, Daniel recognised the role of SSI as a context for other components (e.g., CK and NOS). Yet, it is noteworthy that he did not seem to recognise how CK and NOS facilitated the negotiation of SSI. In both cases, participants did not seem to come to grips of addressing the *interrelationship* between SSI and other components (e.g., CK and NOS) in the curriculum. Both Rick and Daniel belonged to the *Vision I-oriented* group, yet Lillian, who belonged to the *Vision II-oriented* group, had a different opinion. She aptly pointed out that 'SSI serves as the context to help students learn about NOS [...] However, this is not a *unidirectional* facilitation but a *bi-directional* interaction [...] meaning that one always provides opportunity to better understand the other'. This argument suggested an association between understanding the interrelationship between SSI and other components in the curriculum (e.g., NOS) and prioritising SSI teaching in the science curriculum.

3.4.3 Key Learning Moments

The two SSI-related Modules involved classroom activities identified as key learning moments for participants as individuals (See Table 3.7).

First, for *SSI from a learner perspective*, participants perceived that understanding the *nature of SSI* was important to their learning journey. By understanding the complexity of SSI, Morris became more aware of the challenges associated with SSI learning and teaching. This drew his attention to the need for more thoughtful planning for SSI teaching:

> ... the differences between social issues and SSI were identified through classroom activities, which will help me explain the characteristics and importance of SSI to students when

Table 3.7 Summary of key learning moments

Categories	Topics	Classroom activities generating key learning moments
Module 2: SSI from a learner perspective	Nature of SSI: SSI vs LS; SSI vs pseudoscience	Interactive discussion
	Socio-scientific reasoning	Emergent graphical interpretation
	Synthesis of key ideas and practices	Peer debate
Module 3: SSI from a teacher perspective	Pedagogy of SSI	Video analysis

teaching SSI in the future. I realised that the complexity of SSI creates barriers not only for students learning about SSI, but also for teachers discussing SSI. Therefore, I will need to think of ways to effectively teach and address SSI in science class to facilitate SSI learning for students…

Second, participants appreciated the class discussion on the differences between SSI teaching in science class and LS class, which offered them a platform to reflect on these differences. Understanding these differences helped participants build their identity as science teachers and recognise their unique role in student learning, as illustrated by the following excerpt:

the class discussion on mad cow disease reminded me of the complexity and difficulty of talking about SSI and prompted me to think about how to approach the SSI discussion in my science class. Comparing the approach to SSI in LS class and science class, the main difference is that we use rigorous scientific reasoning and apply scientific knowledge more extensively in science classes. This difference is what I will emphasise in my future teaching (Bianca).

Third, participants valued the discussion on pseudoscience. Although science majors, it did not occur to the participants that some people could actually consider global warming as a fallacy, as indicated by Bianca's statement excerpt: '… we watched a video on the credibility of global warming […] I was astonished that some people actually argue that global warming is a fallacy, while for me global warming is an absolute truth'. This warned Bianca that students might not always share the same beliefs as most scientists, which made her aware of the need to explicitly explain to students the differences between scientific claims and pseudo-scientific claims.

Fourth, participants identified the emergent graphical interpretation of socio-scientific reasoning as a key learning experience, as indicated in the following excerpt from Bianca's statement:

…there was a noticeable moment when some of us had problems with the interpretation of the graph on global energy. Never before did I have difficulty reading graphs, because my teachers always analysed the information for us and gave us the essential ideas. Therefore, I realised that I was too comfortable with my current method of dissemination. If I want to develop students' critical thinking and problem-solving skills, then I will have to think more about how to use the materials.

Graphs without full information posed challenges in terms of their interpretation and prompted Bianca to think about how to use graphs in her teaching, especially to develop students' critical thinking and problem-solving skills.

Fifth, for peer debate, consider the following statement written by Sam in his reflective journal:

Before this Module, I had *little incentive to spend time discussing SSI* during my lessons because SSI was rarely assessed in the public examination. Triggered by the in-class debate experience, I found that *participating in SSI teaching is very different from acquiring scientific knowledge* […] I believe that discussing SSI can develop students' scientific literacy. On the one hand, students may have *a better understanding of the scientific concepts* involved by learning from nanomaterials. On the other hand, students may *acquire the ability to interpret and evaluate scientific information* to make an informed decision.

As a result of his participation in the debate, Sam realised that learning about SSI not only promoted the acquisition of CK, but also the skills of interpreting and evaluating scientific information. After the debate, he was motivated to spend time discussing SSI with his students. In other words, participating in SSI debates as learners gave participants first-hand experience in SSI learning, especially in terms of potential learning outcomes.

During the Module on *SSI from a teacher perspective,* video analysis was identified as a learning activity creating key learning moments, as illustrated by Lillian's statement:

> I found this Module very useful because it allowed me to understand the importance of teachers to facilitate student learning and develop their reasoning skills, which will be beneficial throughout their lives to make better judgements and decisions in the future.

The video analysis of SSI teaching emphasised various SSI teaching strategies, but also explicitly identified the goals of SSI teaching (e.g., developing reasoning skills).

These results highlighted key learning experiences from the perspective of participants, with some relevant experiences for their change of beliefs about SSI teaching (e.g., acknowledging the role of SSI teaching in skill development) and others with implications for their implementation of SSI teaching (e.g., recognising the complex nature of SSI).

3.5 Implications for Teaching and Research

The following section summarises the key findings of our exploratory attempt to promote SSI teaching in Hong Kong classrooms by addressing PSTs' beliefs about SSI teaching in a teacher education course:

- Participants' beliefs about SSI teaching shifted from an *instrumental view* to one that went *beyond an instrumental view* after the course;
- The *Vision II-oriented* group was generally better able to identify citizenship education as one of the 'good reasons' for prioritising SSI teaching in the science curriculum;
- The *Vision I-oriented* group tended not to give priority to SSI teaching for the following reasons: (1) the complexity of SSI teaching; (2) the shared curricular objectives of other subjects; and, (3) the subsidiary role of SSI to CK and NOS;
- The following key learning experiences were identified as essential: (1) in-class discussion about the nature of SSI; (2) emergent graphical interpretation for SSI reasoning; (3) peer debate for synthesising key ideas and practices; and (4) video analysis workshop on SSI pedagogy.

Our analysis of why PSTs do not prioritise SSI teaching helps us understand why they struggle to recognise the importance of SSI teaching in the curriculum. One possible reason is their limited understanding of the relationship between SSI, CK and NOS. For instance, some PSTs held a belief indicative of a *unidirectional* relationship between SSI, CK and NOS (e.g., SSI as a vehicle for teaching CK and NOS), ignoring the effects of CK and NOS on SSI teaching (e.g., CK and NOS as conceptual tools to support student argumentation and decision-making related to SSI). Although previous studies have demonstrated the close *interrelationship* between SSI and NOS (Karisan and Zeidler 2017) and between SSI and CK (Sadler and Zeidler 2005), our results highlighted the importance of addressing the *interrelationship* between SSI, CK and NOS in initial teacher education courses aimed at preparing PSTs for SSI teaching. Otherwise, PSTs may continue to see SSI as subsidiary to CK and NOS, affecting in turn whether and how SSI is implemented in science classrooms.

Although our results were based on a single secondary science education programme, limiting their generalisability to other contexts, our findings and insights are of direct concern and relevance to science educators working with PSTs to promote SSI teaching. The results inspired us, as teacher educators, to think critically about course design to identify areas that could be improved for other teacher educators. Similar to NOS teaching, our data led us to speculate that an explicit approach may be more effective in developing a *beyond an instrumental view* on SSI teaching. For instance, engaging PSTs in debates on *why* SSI should be taught will allow us to better identify their intuitive views on SSI teaching. This explicit reflective instruction may also draw PSTs' attention to the *interrelationship* between SSI, NOS and CK. As the current study focused only on PSTs' beliefs about SSI teaching, future studies might usefully explore how PSTs translate their beliefs into their instructional practices by focusing on how they plan and implement SSI teaching in their classroom instruction.

References

Abd-El-Khalick, F., & Lederman, N. G. (2000). Improving science teachers' conceptions of nature of science: A critical review of the literature. *International Journal of Science Education, 22*(7), 665–701.

Abell, S. K., & Bryan, L. A. (1997). Reconceptualizing the elementary science methods course using a reflection orientation. *Journal of Science Teacher Education, 8*(3), 153–166.

Bryan, L. A., & Atwater, M. M. (2002). Teacher beliefs and cultural models: A challenge for science teacher preparation programs. *Science Education, 86*(6), 821–839.

Bryce, T., & Gray, D. (2004). Tough acts to follow: The challenges to science teachers presented by biotechnological progress. *International Journal of Science Education, 26*(6), 717–733.

Bybee, R., McCrae, B., & Laurie, R. (2009). PISA 2006: An assessment of scientific literacy. *Journal of Research in Science Teaching, 46*(8), 865–883.

Creswell, J. W. (2007). *Qualitative inquiry and research design: Choosing among five approaches* (2nd ed.). Thousand Oaks: Sage.

Creswell, J. W. (2008). *Educational research: Planning, conducting, and evaluating quantitative and qualitative research* (3rd ed.). Upper Saddle River: Pearson Education.

Curriculum Development Council. (2017). *Science education key learning area curriculum guide (primary 1-secondary 6)*. Hong Kong: Govt. Logistics Dept.

Curriculum Development Council & Hong Kong Examinations and Assessment Authority. (2015). *Liberal studies curriculum and assessment guide (secondary 4–6)*. Hong Kong: Govt. Logistics Dept.

Demirdöğen, B., Hanuscin, D. L., Uzuntiryaki-Kondakci, E., & Köseoğlu, F. (2016). Development and nature of preservice chemistry teachers' pedagogical content knowledge for nature of science. *Research in Science Education, 46*(4), 575–612.

Hodson, D. (2003). Time for action: Science education for an alternative future. *International Journal of Science Education, 25*(6), 645–670.

Holbrook, J. (2008). Introduction to the special issue of science education international devoted to PARSEL. *Science Education International, 19*(3), 257–266.

Hume, A. (2009). Authentic scientific inquiry and school science. *Teaching Science, 55*(2), 35–41.

Jarman, R., & McClune, B. (2007). *Developing scientific literacy: Using news media in the classroom*. Berkshire: McGraw-Hill Education.

Kara, Y. (2012). Pre-service biology teachers' perceptions on the instruction of socio-scientific issues in the curriculum. *European Journal of Teacher Education, 35*(1), 111–129.

Karisan, D., & Zeidler, D. L. (2017). Contextualization of nature of science within the socio-scientific issues framework: A review of research. *International Journal of Education in Mathematics. Science and Technology, 5*(2), 139–152.

Khishfe, R. (2012). Nature of science and decision-making. *International Journal of Science Education, 34*(1), 67–100.

King, K., Shumow, L., & Lietz, S. (2001). Science education in an urban elementary school: Case studies of teacher beliefs and classroom practices. *Science Education, 85*(2), 89–110.

Lee, H., & Witz, K. (2009). Science teachers' inspiration for teaching socioscientific issues (SSI): Disconnection with reform efforts. *International Journal of Science Education, 31*(7), 931–960.

Lee, H., Abd-El-Khalick, F., & Choi, K. (2006). Korean science teachers' perceptions of the introduction of socio-scientific issues into the science curriculum. *Canadian Journal of Math, Science & Technology Education, 6*(2), 97–117.

Lee, H., Yoo, J., Choi, K., Kim, S. W., Krajcik, J., Herman, B. C., & Zeidler, D. L. (2013). Socioscientific issues as a vehicle for promoting character and values for global citizens. *International Journal of Science Education, 35*(12), 2079–2113.

Leung, J. S. C., Wong, A. S. L., & Yung, B. H. W. (2015). Understandings of nature of science and multiple perspective evaluation of science news by non-science majors. *Science & Education, 24*(7), 887–912.

Lumpe, A. T., Haney, J. J., & Czerniak, C. M. (1998). Science teacher beliefs and intentions to implement science-technology-society (STS) in the classroom. *Journal of Science Teacher Education, 9*(1), 1–24.

McComas, W. F. (1998). The principal elements of the nature of science: Dispelling the myths. In W. F. McComas (Ed.), *The nature of science in science education* (pp. 53–70). Dordrecht: Kluwer.

Merriam, S. B. (2009). Qualitative case study research. In S. B. Merriam (Ed.), *Qualitative research: A guide to design and implementation* (pp. 39–54). San Francisco: Jossey-Bass.

Pedersen, J. E., & Totten, S. (2001). Beliefs of science teachers toward the teaching of science/technological/social issues: Are we addressing national standards? *Bulletin of Science, Technology & Society, 21*(5), 376–393.

Ratcliffe, M., & Grace, M. (2003). *Science education for citizenship: Teaching socioscientific issues*. Maidenhead: Open University Press.

Roberts, D. (2007). Scientific literacy/science literacy. In S. K. Abell & N. G. Lederman (Eds.), *Handbook of research on science education* (pp. 729–780). Mahwah: Lawrence Erlbaum Associates.

Sadler, T. D. (2004). Informal reasoning regarding socioscientific issues: A critical review of research. *Journal of Research in Science Teaching, 41*(5), 513–536.

Sadler, T. D. (2011). Situating socio-scientific issues in classrooms as a means of achieving goals of science education. In T. D. Sadler (Ed.), *Socio-scientific issues in the classroom: Teaching, learning and research* (pp. 1–9). Dordrecht: Springer.

Sadler, T. D., & Zeidler, D. L. (2005). The significance of content knowledge for informal reasoning regarding socioscientific issues: Applying genetics knowledge to genetic engineering issues. *Science Education, 89*(1), 71–93.

Sadler, T. D., Chambers, F. W., & Zeidler, D. L. (2004). Student conceptualizations of the nature of science in response to a socioscientific issue. *International Journal of Science Education, 26*(4), 387–409.

Sadler, T. D., Romine, W. L., & Topçu, M. S. (2016). Learning science content through socio-scientific issues-based instruction: A multi-level assessment study. *International Journal of Science Education, 38*(10), 1622–1635.

Sadler, T. D., Foulk, J. A., & Friedrichsen, P. J. (2017). Evolution of a model for socio-scientific issue teaching and learning. *International Journal of Education in Mathematics, Science and Technology, 5*(2), 75–87.

Tidemand, S., & Nielsen, J. A. (2017). The role of socioscientific issues in biology teaching: From the perspective of teachers. *International Journal of Science Education, 39*(1), 44–61.

Tosunoglu, C. H., & Lederman, N. G. (2016). *The development of an instrument for assessing pedagogical content knowledge for Socioscientific Knowledge (PCK-SSI).* Paper presented at the 2016 annual meeting of NARST, Baltimore, MA.

Vesterinen, V. M., Tolppanen, S., & Aksela, M. (2016). Toward citizenship science education: What students do to make the world a better place? *International Journal of Science Education, 38*(1), 30–50.

Yip, V. W. Y., Chan, K. K. H., Yung, B. H. W., & Lai, C. (2018). Preparing teachers for reform-oriented teaching using the thematic approach of re-viewing videos. In L. H. Xu, G. Aranda, & D. Clarke (Eds.), *Video-based research in education: Cross-disciplinary perspectives* (pp. 260–274). New York, NY: Routledge.

Zeidler, D. L. (2014). Socioscientific issues as a curriculum emphasis: Theory, research, and practice. In N. G. Lederman & S. K. Abell (Eds.), *Handbook of research on science education* (Vol. II, pp. 697–726). New York: Routledge.

Zeidler, D. L., Sadler, T. D., Simmons, M. L., & Howes, E. V. (2005). Beyond STS: A research-based framework for socioscientific issues education. *Science Education, 89*(3), 357–377.

Zeidler, D. L., Applebaum, S. M., & Sadler, T. D. (2011). Enacting a socioscientific issues classroom: Transformative transformations. In T. D. Sadler (Ed.), *Socio-scientific issues in the classroom: Teaching, learning and research* (pp. 277–305). Dordrecht: Springer.

Jessica S. C. Leung is an Assistant Professor in the Division of Mathematics and Science Education, Faculty of Education, the University of Hong Kong (HKU). Before that she was a Lecturer in the Faculty of Science for the same university. She earned her Ph.D. in Science Education, her M.Phil. in Food Science and her B.Sc. at HKU. Leung is currently the coordinator of the Bachelor of Education and Bachelor of Science programme for the university. She teaches courses for pre-service secondary science teachers as well as general education courses with a focus on socio-scientific issues for students from all Faculties. Leung's research has focused on how students evaluate socio-scientific issues and scientific claims reported in the media as well as its relationship with students' understanding of the nature of science.

Ka Lok Wong is Senior Lecturer in the Faculty of Education (Division of Mathematics and Science Education) at The University of Hong Kong. Much of his teaching duties falls in the areas of mathematics education as well as the training of teachers in liberal studies which cover an area of study called *Science, Technology and the Environment*. He also engages himself in a range of educational communities and professional services, including the presidency of Hong Kong

Association for Mathematics Education and leadership in various government-funded projects supporting the learning and teaching of mathematics in schools. In addition to his involvement in PISA for more than a decade, his research interests include the development of mathematics curriculum, history and philosophy of mathematics, use of information technology in mathematics education, and teacher-student interaction in the mathematics classroom.

Kennedy K. H. Chan is an Assistant Professor at the University of Hong Kong (HKU). He received his BSc and MPhil degrees in Science from HKU. Before pursing his Ph.D. studies at the same university, he worked as a secondary school science teacher in local secondary schools. His research interests include pedagogical content knowledge (PCK), teacher noticing, using videos to promote teacher learning. He is a recipient of various scholarships and awards including Hong Kong Ph.D. Fellowship, Sir Edward Youde Memorial Fellowship (For Postgraduate Research Students) and International Science Education Conference (ISEC)-Springer Best Paper Award (Student Category). He was an invited participant of PCK summit II held in the Netherlands.

Chapter 4
Socio-Scientific Inquiry-Based Learning: Possibilities and Challenges for Teacher Education

Ruth Amos, Marie-Christine Knippels, and Ralph Levinson

4.1 Introduction

One of the characteristics of school science curricula over the years is that they tend to privilege canonical science, i.e., an emphasis on scientific laws, theories, facts and principles over the social context of science (Millar and Osborne 1998); what Roberts (2011) refers to as a Vision I view of scientific literacy. Although there have been periodic reforms on the social context of science or citizen science, for example in the UK, twenty-first century Science (Millar 2006), PLON in the Netherlands (Wierstra and Wubbels 1994) and in the US, SEPUP, (Koker 1996) teachers have nonetheless found it difficult to implement teaching of socio-scientific issues (Day and Bryce 2011; Levinson and Turner 2001) within the constraints of a Vision I curriculum. With the goal of moving beyond such constraints, in this chapter, we reflect on our experiences as teacher educators developing a pedagogical approach for pre-service science teachers (PSTs) teaching through socially responsible inquiry embedded in contemporary socio-scientific issues. We outline the comparative requirements of the national curricula of such learning in school science (post-2011) in England, the Netherlands and Sweden. Finally, we report on three short case studies in which PSTs designed and taught activities using this approach, with a view to highlighting the opportunities and challenges they encountered.

R. Amos · R. Levinson (✉)
University College London, Institute of Education, London, UK
e-mail: ruth.amos@ucl.ac.uk; r.levinson@ucl.ac.uk

M.-C. Knippels
Freudenthal Institute, Utrecht University, Utrecht, The Netherlands
e-mail: M.C.P.J.Knippels@uu.nl

M. Evagorou et al. (eds.), *Science Teacher Education for Responsible Citizenship*, Contemporary Trends and Issues in Science Education 52,
https://doi.org/10.1007/978-3-030-40229-7_4

In terms of a Vision I curriculum, one aspect of the problem is that to fulfil curriculum and examination requirements teachers have understandably tended to focus on core, substantive concepts where social and ethical issues have become an illustrative bolt-on (Levinson and Turner 2001) rather than a source of controversy where both descriptive and normative considerations can be brought to discussion (Kolstø 2001; Sadler 2009; Boerwinkel et al. 2014). The typical Vision I curriculum is based predominantly on covering laws such as Newton's Laws of Motion, the Law of Falling Bodies and Boyle's Gas Laws, where facts are dichotomised from values, so integrating scientific, social and political issues into a teaching scheme is always going to be a tough pedagogical challenge (Dawson 2000; Ryder 2002). In recent years, socio-scientific issues feature less in the National Curriculum in science in the UK with more emphasis on facts and principles. Tim Oates, an educational advisor to the UK Conservative government, commenting on a review of the National Curriculum illustrates this point: 'We have believed that we need to keep the National Curriculum up to date with topical issues but oxidation and gravity don't date …we are taking it back to the core stuff.' (www.guardian.co.uk/education/2011/jun/12/climate-change-curriculum-government-adviser).

Yet, one of the main contemporary educational programmes in Europe is to gear democratic citizenship with responsible research and innovation (RRI) in science and technology (http://www.euroscientist.com/towards-responsible-research-innovation/) (Sutcliffe 2011). Its main aims are that science education should align scientific research both with and for society (Owen et al. 2012), that the products of science and technology both meet standards of social desirability, ethical acceptability and sustainability (Von Schomberg 2014) and public participation in this process takes place both upstream and downstream throughout the research and development process. Socio-scientific issues (SSI) have the potential to give students insights into such processes (Lee et al. 2013; Gutierez 2015). Technologies such as nanotechnology and robotics are clear targets of this process. Ravetz (2004), in depicting post-normal science, identified technologies such as nanotechnology, robotics and artificial intelligence as those where decision stakes and social uncertainties are high, presenting potential unknown hazards. Nanomaterials have the potential to solve many problems in society but such small particles present health risks (Shatkin 2013). Robots meet some economic and social needs, for example, performing repetitive human tasks, and undergoing complex manoeuvres beyond the possibilities of human certainty, but they also risk increasing unemployment, redundancy and social unrest (http://www.fusbp.com/wp-content/uploads/2010/07/AI-and-Robotics-Impact-on-Future-Pew-Survey.pdf).

Promoting Attainment of Responsible Research and Innovation in Science Education (PARRISE) was a European Union (EU) project consisting of a consortium of partners from 18 universities and 11 different countries. The overarching objective of PARRISE (www.parrise.eu) was to elaborate pedagogies which bring together, under the umbrella of RRI, three supporting and mutually interactive pillars, namely, inquiry based science education (IBSE) and learning of socio-scientific issues (SSI) incorporating Citizenship Education (CE); and to develop

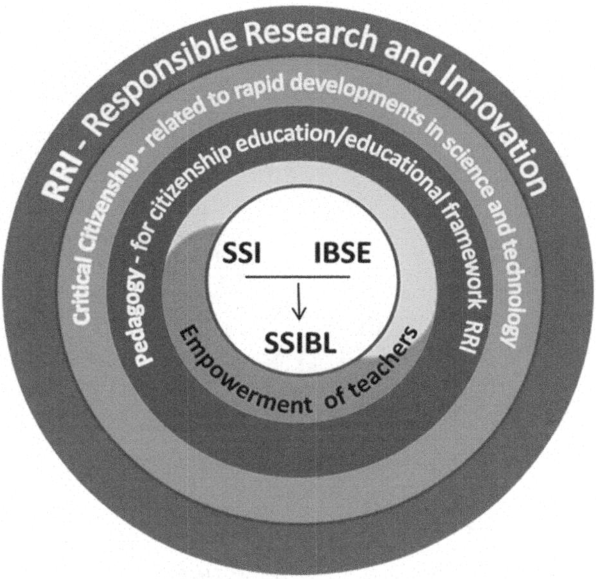

Fig. 4.1 Promoting Attainment of Responsible Research and Innovation in Science Education (PARRISE): Main components

these pedagogic competences among science teachers (Knippels and van Dam 2017). In this chapter, we highlight our work in the project with pre-service science teachers (PSTs), looking for opportunities to adopt an 'it matters' approach towards science for and with society (Fig. 4.1).

RRI is conceived as the overarching context, with CE, IBSE and SSI forming three pillars of the pedagogy. These three pillars are interrelated. SSIs incorporate the pre-requisites for inquiry: the contextualised knowledge; the relevant socio-scientific scenarios; and, the strategies to stimulate interest (Sadler 2009). IBSE is the process of collecting evidence to find solutions to authentic questions arising from student interest. This is a complex process for PSTs to learn to manage in their training year (Roehrig and Luft 2004). For example, evidence collected and reflected upon might influence re-formulation of the original inquiry. The implications of this process are considered in relation to the personal, social and global aspects of the inquiry, that is, issues which affect the individual, such as recycling household waste, need to be considered also in broader terms of the energy costs of recycling and the effects of recycling in a global context.

Democratic and participative processes through which the inquiry is carried out involve open and critical discussion. Finally, the evidence contributes towards relevant action taken to address the original question. Since RRI itself is a socialisation process into doing research, we developed an educational framework – the PARRISE model – by iterative modification in two cycles by the 18 teacher training institutions through feedback from science teachers and teacher educators taking part in the project. The model was modified to reflect the RRI context: finding a solution to socio-scientific problems through research-based inquiry. We refer to this overall

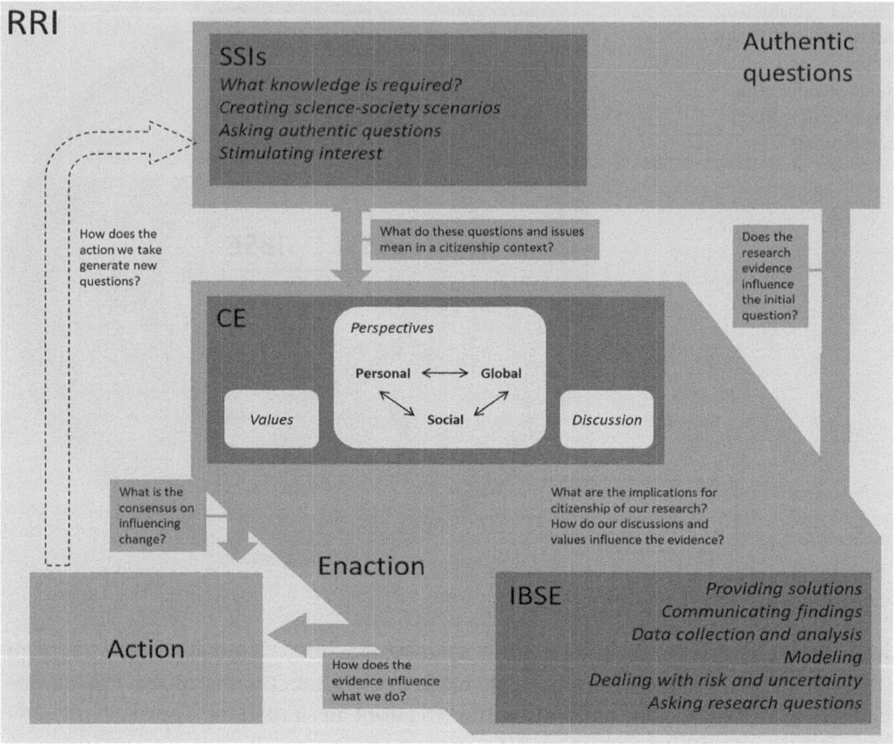

Fig. 4.2 Representation of the SSIBL approach. The approach is embedded within the overall context of RRI

process of inquiry in finding solutions to authentic socio-scientific questions as ***Socio-Scientific Inquiry-Based Learning*** (**SSIBL**). Figure 4.2 represents the learning model of SSIBL. It is directed at all school age groups from 5 to 19 although our focus in this chapter is on mid-range groups between the ages 11–15. Our refined model for teachers (Levinson et al. 2017) is available on the project website (www.parrise.eu).

The SSIBL approach involves three main stages, although the stages are not necessarily sequential:

(a) Raising an authentic research-based question about a socio-scientific issue;
(b) Carrying out research-based inquiry on the question to enact change (such a process might involve carrying out experiments and/or surveys, analysing data, collaborating with others);
(c) Finding a solution (this involves communicating results based on evidence from research, convincing others of the necessity for change, particularly those who have influence to carry out the change).

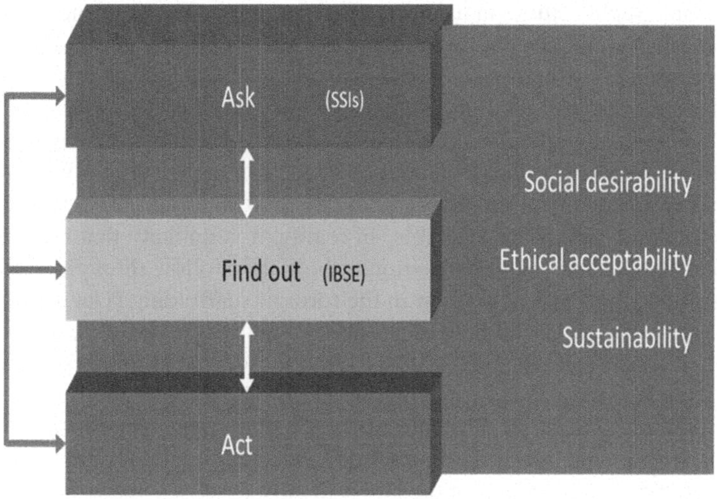

Fig. 4.3 Summary of SSIBL

Together these can be summarised for teachers and learners as Ask- Find Out -Act as illustrated in Fig. 4.3.

4.1.1 Raising Authentic Questions – Ask

There are two main interpretations of authenticity, one related to the practice of scientists (Chinn and Malhotra 2002), the other – which is the subject of our chapter – is that the problem arises as much as possible from the 'interests, perspectives, desires, and needs of the students' (Buxton 2006, p. 701) rather than being directed by the teacher. These types of inquiry are more likely to be achieved in non-formal education settings given the contemporary emphasis in many post-industrial countries of short-term objectives for schools meeting externally imposed standards for accountability measures (Hargreaves and Shirley 2009). Brickhouse (2011) reports such effects, preferring non-formal approaches when attempting to introduce inquiry activities she had originally run in out-of-school environments. Roth (1997), drawing on situated learning, stresses the socio-cultural nature of inquiry, where practices are shared within a community of inquiry, common goals and purposes are negotiated, and where knowledge can be used in different contexts. Problems are open-ended and loosely framed and, in improvising collectively, produce and act on failure as part of this struggle.

In an ethnographic study on teaching contextually authentic science inquiry in a struggling urban elementary school in the US, Buxton (2006) identified the following prerequisites: a willingness of teachers to adopt a model of curriculum that links to issues of personal and local interest, as well as taking up the opportunity to use 'teachable moments' to pursue an inquiry. An assessment model which incorporates whole school planning (Harris and Ratcliffe 2005), and paying attention to students' choices in documenting their learning are also important. However, resources and opportunities need to be available. In reality, it is unlikely that many students will raise spontaneous research questions which they follow through on their own without some authoritative direction in the form of scaffolding (Davis and Miyake 2004). Roth (1997) also demonstrates how effectively students can achieve in school when set unstructured tasks, nonetheless these were initially set by researchers and teachers.

How authentic a question is for students is a matter of degree: one student with a deep interest in a topic might persuade less interested peers to collaborate without any real commitment on behalf of the latter; in most cases it is likely that the teacher will have some input (Levinson and the PARRISE consortium 2017) but how the teacher enables student interest and engagement will follow a spectrum of authenticity.

4.1.2 An Inquiry-Based Approach – Find Out

Adopting an inquiry-based approach to address authentic socio-scientific questions is an adaptation of a focus on the creativity and possibilities of inquiry in EU policy for science education (Rocard et al. 2007) and has a recent history of evidenced advocacy (Gormally et al. 2009; Wilson et al. 2010). However, what is meant by inquiry-based approaches has a range of interpretations. Abd-El-Khalick et al. (2004) distinguish between inquiry as: means – inquiry *in* science – as a way of teaching to help students understand science content, inquiry as ends, or inquiry *about* science involving students learning about epistemic aspects of science practice and development of knowledge, as well as inquiry skills such as generating research questions. In the context of our study and PARRISE more generally, the emphasis is on 'inquiry as ends' although such an approach does not preclude gaining scientific knowledge. Lederman et al. (2014) point out the distinction between inquiry for scientific literacy in making informed decisions where the emphasis is on how science is done and how knowledge is produced, distributed and evaluated rather than necessarily performing inquiries. They lay stress, however, on the teacher promoting explicit reflection on knowledge *about* inquiry. But it should be stressed that these critiques are aimed specifically at scientific inquiry while SSIBL is inter-disciplinary.

In the terms used in this study, inquiry is used in a broader interdisciplinary manner to indicate the processes of finding an answer or answers to an open socio-scientific question involving reflection and primary data collection in the form of evidence. In the context of SSIBL, the focus of research and data collection goes beyond canonical substantive Vision I Science, and might comprise collection of data through social research, i.e., surveys and interviews, as well as scientific experiments. We recognise that this interpretation is loose but also allows us to refine any depictions of inquiry through accumulating experience and theorisation of research.

4.1.3 Finding a Solution – Act

Finding a solution entails the enactment of change. A question being authentic presupposes that the object of interest is perceived as unsatisfactory and needs to be changed. Much of the research on student involvement in SSIs has involved asking questions about relevant problems (Ekborg et al. 2013; Boerwinkel et al. 2011) and taking up, for example, role-play positions (Simonneaux 2001), but the findings have not necessarily been translated into affecting material change. Our aim was for science teachers to support students in considering personal action resulting from their research and raised awareness.

Bouillon and Gomez (2001) reported on the activities of the Chicago River Project in which elementary school students along with local community agencies were involved in cleaning up the banks of a polluted river. Their programme of 'connected science' 'uses real-world problems and school-community partnerships as contextual scaffolds to bridge [...] diverse funds of knowledge' (p. 895). What enabled action in an open-ended inquiry with messy data were interdisciplinary collaborations, relevance both to the curriculum and to the living context of the children's lives. The success of action can be attributed to a genuine partnership between community agencies and the school, all who shared a commitment, albeit through diverse tactics, to solve the problem. Roth and Lee (2004) have shown that multi-party inquiries involving school students in environmental problems such as waterway pollution raise political issues about how change resulting from findings can be enacted. Calabrese Barton and O'Neill (2008) demonstrate how school students focusing on place and identity raise controversial questions about curriculum change. Through Research-informed Action (RiA), Krstovic (2014) has enabled high school students to raise awareness about powerful corporations, using evidence-based research to produce videos, brochures, posters, devising new labels for water bottles, and class presentations to lobby for action. In preparing for RiAs, Krstovic suggests that students are encouraged to explicate the issue, identify what they already know and what they would need to learn, propose an idea for collecting social data, such as surveys, and establishing reasons to take action on the issue.

4.1.3.1 Adopting the SSIBL Approach

Teachers are often constrained by the practicalities of curriculum space. First we examine the opportunities and constraints for a SSIBL approach in England, the Netherlands, and Sweden (countries representing consortia in PARRISE) by comparing their curricula. Secondly we analyse SSIBL activities carried out by three pre-service teachers in different London schools to identify their effectiveness in terms of opportunities and constraints presented to teachers. We then appraise the process of pedagogy and research, proposing possible ways forward.

4.1.3.2 SSIBL and National Curricula

A SSIBL type approach has been implemented in teacher professional development sessions in 11 European countries from 2014 to 2018 in the context of the PARRISE project. Diverse training modules were used that varied in length from one or two to multiple sessions that took place over the time span of a few months. This variety is partly due to differences in national contexts, curriculum opportunities and teachers' (educators) preferences. In some countries the curricula are flexible enough to allow for experimentation and innovation, while in others these kinds of opportunities are more limited. Some teacher training institutions introduced the SSIBL framework explicitly from the outset (a more deductive approach) whereas others introduced and explored it gradually (a more inductive approach). But what all teacher professional development modules had in common was that they were based on opportunities for experiential learning and collaborative practice, including cycles of SSIBL lesson design, implementation and reflection. This allowed pre-service and experienced science teachers to contribute to the refinement of the SSIBL educational model while developing their practice, alongside their supporting teacher educators.

To exemplify these differences, we briefly compare the opportunities and constraints in the curriculum for a SSIBL type approach in three countries: England, the Netherlands and Sweden. Then we discuss concrete case studies of pre-service teachers implementing SSIBL in England.

What stands out is that for all three countries the SSIBL-related requirements are formulated predominantly as overarching general competences. However, the biology and chemistry curricula in the Netherlands (https://www.examenblad.nl/) have content specific requirements related to sustainability such as human influence on the system Earth, sustainable production processes and energy preservation. Likewise, the biology, chemistry and physics curricula in Sweden (https://www.skolverket.se/publikationer) have similar requirements to examine the impact of people on nature, locally and globally, and opportunities for consumers and citizens of society to contribute to sustainable development. In physics, for example, students are taught about electricity production, distribution and use in society, including the supply and use of energy historically and currently, as well as possibilities and limitations for the future. The chemistry curriculum explores people's use of energy and natural resources, locally and globally, in the context of sustainable development. Additionally, students are taught about chemical processes in the manufacture and recycling of metals, paper and plastics, and processes for purifying drinking water and waste water, locally and globally. The science curriculum for England (https://www.gov.uk/government/publications/national-curriculum-in-england-science-programmes-of-study) has only very limited reference to science and society or socio-scientific issues; students are required to study, for example, topics such as healthy living, energy resources and recycling briefly but more general competences such as opinion-forming or communication and information skills are not mentioned. Changes to the English national curriculum for science in 2014 reduced the emphasis on 'application and implications of scientific developments'. The focus shifted to scientific inquiry, experimental skills, concepts and facts and so is in that sense a more Vision I related emphasis of the curriculum.

All three curricula include scientific inquiry skills, in other words students should be able to formulate and analyse questions, carry out research, interpret data and draw conclusions from the research results. The Swedish curriculum explicitly refers to analysing and seeking answers to topic-related questions as well as identifying, formulating and solving problems. It does not state specifically that students should raise topic- or societal-related (authentic) questions, but this curriculum requirement does give this opportunity related to raising questions in the SSIBL approach. Moreover, it refers to the importance of science knowledge for individuals and society and that students should be able to use this knowledge to communicate as well as to review and use information. This is an important skill in the inquiry-based stage of the SSIBL approach in which the trustworthiness of different information sources (e.g., internet, stakeholders) should be evaluated.

The Dutch science curriculum also provides opportunities for a SSIBL approach. It includes explicitly that students should be able to give reasoned judgements about situations in nature or in technical applications, and that students should be able to distinguish between scientific arguments, normative social considerations and personal views. In the research phase of SSIBL (*Find out*), inquiring about and weighing up different perspectives (e.g., local, global, personal) are important aspects as well as dealing with scientific data and arguments, e.g., trustworthiness of sources, uncertainty of findings, communication. The biology and chemistry curricula have explicit content specific requirements related to sustainability, providing opportunities to link SSIBL to the regular 'core curriculum', such as it might be experienced by science teachers, since the more general competences are assessed less at the end of secondary education.

It can be seen that there are content requirements, particularly in the Dutch and Swedish curricula, which are consistent with a SSIBL approach. They allow for the first three stages of Hodson's (2003) levels of issues-based approaches, namely appreciating the social impact of science and technology, recognising political links with science and society, and developing value positions but omit both the last stage of preparing for and taking action, and give no clear guidance for promoting authentic questions. In these three national jurisdictions, there is limited but varied scope for teachers to build on content in the curriculum. We now look at pre-service teachers in the context of the English national curriculum – since the current English curriculum portrays the most Vision I related approach – to gain insight into the opportunities afforded by SSIBL as well as the constraints drawing on the SSIBL model. We want to highlight examples showing a range of experiences from the PSTs' perspectives in developing SSIBL activities, where they recorded detailed accounts through the 'design-enactment-reflection' stages (Schön 1983). As the three examples are in a similar field, we were able to compare and contrast specific aspects of the SSIBL model adopted by these PSTs.

4.2 Method

To illustrate some *particular* features of the implementation of SSIBL, we report on three case studies (Stake 1995). The case is the pre-service teacher (PST), class and SSIBL activity. They are cases because the schools were demographically and socially distinct, the teachers' backgrounds were different and the schools' systems of support varied. We therefore were focused on different contexts rather than making generalisations about SSIBL at this stage.

The cases come from 2 years of the PARRISE project, when two cohorts of pre-service science teachers (173 in total) worked to plan and teach SSIBL activities in their practicum schools. PSTs were first inducted into the process of SSIBL through a number of interactive scenarios. Resources encouraged PSTs to brainstorm questions which might arise from the stimulus material. These consisted of artefacts and reports such as articles in the media about e-cigarettes and their efficacy in changing smoking habits among female school pupils, artefacts which linked together various materials in environmental depredation, for example, pictures of bauxite excavation, the wrappers on chocolate products, and flooding caused by alkaline purification of alumina (all involved in the extraction, manufacture and uses of aluminium) (Levinson 2009). The purposes here were to:

(a) encourage PSTs to reflect on the use of stimulus materials and to ask authentic questions, and how they might use or adapt such materials through their teaching;
(b) identify areas of the curriculum which could be adapted for SSIBL activities;
(c) co-design SSIBL activities which they could enact in their teaching practices (Kyza and Nicolaidou 2017);
(d) generate written reflections to share with colleagues after practice.

During the training year, PSTs co-planned and co-designed, with various degrees of input from experienced teacher mentors, SSIBL activities, resulting in implementation and reflection. The approach allows for cycles of *design – enactment/implementation – reflection,* (Schön 1983) which took the form of co-planning in university sessions, implementing lesson plans in school, then reflecting on their own evaluations and those of their science subject mentors (Voogt et al. 2011). This also encouraged teachers to promote further change and possibilities in developing reform (Kyza and Nicolaidou 2017) within the context of their own practice, such as integrating and disseminating SSIBL in their own departments and schools. During their training year, PSTs were encouraged to record and reflect on experiences of designing and implementing (enacting) SSIBL. At the end of the course, three PSTs who had taught a sequence of SSIBL lessons were asked to reflect on their experiences through a group discussion[1]; the main data were produced through thematic analysis of lesson plans, resources and post-lesson written reflections, examining the use of dimensions of the SSIBL model. In the latter, university tutors drew on the plans and reflections to then support PSTs on specific aspects of their SSIBL practice, and to inform the iterative development of the SSIBL model over 4 years.

4.2.1 Pre-service Teacher Cases

The three cases described below are based on self-documented plans, reflections and accounts of SSIBL activities given by three PSTs who focused on issues related to healthy living. Health-related topics were current when they were in school designing SSIBL activities. The data consist of lesson plans and detailed, reflective evaluations, tasks and artefacts, and documents used during the lessons.

4.2.1.1 Martina

This series of lessons for 13–14 year-old students on SSIBL began with a lesson from the topic 'Healthy Living' about the respiratory system, hence they started with a curriculum focus. The students were taught about different parts of the respiratory system that are affected by smoking, and Martina demonstrated the effects of smoking on the lungs by using a smoking machine. Smoke from a cigarette is drawn through a horizontal glass tube containing cotton wool. Students can see the effect of tar depositing on the cotton wool and the residue of the smoke which passes

[1] Discussions between three PSTs and a teacher educator can be seen on YouTube https://www.youtube.com/watch?v=oksKEgBe25A and the PARRISE website: https://www.parrise.eu/videos/

through collects in a conical flask containing water. The students discussed how oxygen was carried to metabolising cells through the surface of the lungs and capillaries, how carbon dioxide was removed and how far the smoking machine can be considered a valid model of the lungs and bronchi.

The next lesson started with some stimulus images including a group of school girls smoking, someone e-smoking and headlines and reports on smoking taken from the media. Martina tried to be as open as possible. After discussion, students suggested the question: 'How come teenagers still smoke when they know the dangers?' Students designed questions with peers for an online survey (using *Survey Monkey*™) to send out to 15 friends or family followed by class evaluation and agreement on which questions to use. Analysis of the lesson plans showed that Martina (and indeed other PSTs) saw question-asking and gathering data as the key phases of implementation ('enaction') but did not think about other aspects such as mapping controversies, comparing stakeholder actions, or other possible scaffolding strategies. During the final lesson they analysed the survey responses and created 'a leaflet to dissuade teenagers (i.e. their peers) from starting smoking'.

Martina reported the students to be 'engaged right from the very beginning'. She needed to support some students using prompts and by modelling questions to help them design the survey. Students focused on governmental and social policy as well as the smoking habits of friends and family. Students questioned the validity of the survey, because they felt respondents had probably not been entirely honest. Martina felt the students needed more time to discuss how they created and analysed the survey. Because of time constraints she did not focus in depth on what creating valid survey questions means with the students. Other teachers in the science department took an interest in the activity and planned to develop it for future years. Martina was encouraged to develop further SSIBL activities during her time at the school.

4.2.1.2 Caitlin

Caitlin followed a similar format in her school for the same age group starting from a lesson on the respiratory system and the effects of smoking, then broadening out to the question 'Why do young people smoke?' Unlike Martina, Caitlin took a more structured and guided approach in setting the question. There were some further differences between her account and Martina's. In contrast to Martina, Caitlin discussed the challenge of collecting data with the students, then asked them to consider in groups the effectiveness of different methods of data collection: a survey questionnaire distributed to a sample population of their peers; observations while watching and recording an individual smoking over a period of time as well as the effects on others; controlled experiments testing medical and psychological indicators of a control and intervention population, and interviews with smokers. After discussions about the merits and problems of each method, the students decided as a class to do a survey. Having done so, they further had to decide on sample size and justify the number of people to be surveyed, the number, type and order of questions, and to spend some time improving draft surveys. While all students were reported to be engaged in Martina's lesson, Caitlin reported some low level disengagement and felt she had spent too much time exploring methods with the class.

Students carried out fewer surveys, only three per person, compared with 15 each in Martina's class. Once the data had been returned, students spent some time discussing how best to present and interpret it based on criteria given by Caitlin and discussed explicitly with the students. They created informative posters, which were put up in classrooms of that year group throughout the school.

4.2.1.3 Andrea

Andrea's inquiry with 11 and 12 year-old pupils on fizzy drinks was specifically suggested by her school science mentor, from observations she had made of school pupils' eating and drinking habits in school, and drawing on contemporary concerns about the diets of young people in the UK in general (http://www.bbc.co.uk/news/health-37511554). Andrea designed a three-lesson sequence for pupils to carry out their own investigations into fizzy drinks: setting up plans for the investigation, sharing results from their designed survey, and designing and presenting posters and results.

In planning for the first lesson, Andrea suspected that 'pupils will have lots of knowledge developed from the media portrayals of drinks but little science to back it up'. In supporting their knowledge and design of surveys, she provided the students with key questions and structured planning sheets. But there was no scaffolding to support students in thinking about different stakeholder perspectives. In her post-lesson reflections, she notes there was a wide variety of engagement and collaborative enactment; some students, however, struggled to think of questions while others were fully involved.

Overall Andrea reports:

> All the students found it easy to choose a drink and appeared to have made a good research and survey plan and split tasks between members of their group during the first lesson. However, it became apparent when students came to writing up their research that whilst some groups had done surveys and research many students had not grasped the scientific concepts and just presented 'facts' about the drink they were researching.

Andrea also reflected:

> None of the students made any conclusions relating to how healthy or unhealthy a drink was, or the maximum quantity you might want to drink, despite discussion in the first lesson about how you could relate sugar content to recommend daily allowances. As such, I felt there was little meaningful outcome in an investigation that could have been quite eye-opening for the students. This suggests that I didn't give the students enough guidance or scaffolding in this area, where I was looking for them to come to some kind of conclusion.

Her honest reflections highlight typical challenges faced by PSTs during their training year when they try to combine different pedagogical strategies and skills together (Roehrig and Luft 2004; Ingersoll and Strong 2011).

4.2.2 Summary of Case Studies

These three lesson sequences exhibit distinct features which help to build up a picture of the challenges and possibilities of SSIBL. It is clear that some induction into the main scientific principles of a topic appears to support student interest and understanding of what is at stake in then carrying out socially-responsible research. In the three cases, somewhat unsurprisingly, students struggled with developing appropriate data collection instruments and in analysing what was produced. Caitlin and Andrea were careful to support students in structuring their surveys but it is doubtful that the data produced had validity in relation to the conclusions generated. This is also the case in Martina's experience. It appears at the very least that students need more support to build skills and knowledge in generating small social surveys and analysing the results. This is unlikely to be achieved through isolated inputs. As Krstovic (2017) reports, detailed, careful and explicit scaffolding is needed to support students' competences in designing surveys and working with correlation studies.

All three PSTs received support from their school science mentors which helped them implement the activities. In Martina's case, the enthusiasm both from her mentor and the school science department in particular made ways of building on Martina's experience and implementing more fruitful activities in the following years possible.

As is common during the training year (Roehrig and Luft 2004), our cases revealed PSTs struggled to fully understand, then design and implement some of the complexities of SSIBL, tending to focus on specifics such as students raising authentic questions and constructing survey research, at the expense of other components of the model. Some members of the PARRISE consortium were able to address this through longer term collaboration between PSTs and experienced teachers, which proved fruitful.

4.3 Discussion

The experiences of the three case study PSTs teaching SSIBL reflect their commitment to adopting a SSIBL approach in their science classrooms, as well as a high level of personal reflection. But, particularly in the cases of Caitlin and Andrea, they identified the complex steps that need to be brought into place before students could use science knowledge (Andrea) or confidently devise surveys (Caitlin). These are aspects of SSIBL that need scaffolding. So, in the first years of teaching, including the pre-service year, there are serious challenges in maintaining the commitment to action-based inquiry, such as SSIBL, as well as supporting the skills and knowledge students need to bring SSIBL to fruition.

Scaffolding is a central pivot to social constructivist learning because it presupposes support can be given at a stage when the student is ready for it and can then be phased out when the learner has acquired the required competence (Singer et al. 2000). The precise nature of the support depends on a range of factors, what needs to be learned, the knowledge and skills the learner already has, the experience they have of the context of learning, the complexity of the concepts and skills to be learned, the knowledge and skills of the teacher. All three cases also tried to focus on issues which PSTs and teachers felt were particularly relevant for the students, rather than more 'global' issues (i.e., climate change, IVF, deforestation, etc) and there is some evidence from all the analysed lesson reflections that these kinds of personally-relevant SSIBL activities for teenagers allow for more engaged learning, and the potential for meaningful action on SSIs.

Given that time and curriculum requirements are common constraints for teachers, we suggest that SSIBL could be fruitfully developed through short-term inquiries. These would have outcomes which could be completed in one or two lessons and carried out mainly within school. Longer-term SSIBL could go beyond this timeframe and include external agencies. Examples of short-term and/or relevant SSIBL issues used by PSTs in the PARRISE consortium countries include:

- Designing a healthy diet for the school canteen;
- Reducing the use of plastic drink bottles in schools;
- Improving recycling in school by researching the logistics of bringing recycling bins into classrooms;
- Reducing water consumption;
- Writing a plea against the firm that produces C8 (similar to Teflon™, carcinogen) and dumps waste in the local river;
- Tesla: good for the environment or not? A critical view on news articles;
- Writing a report about sugar content and discussing sugar tax, in context of diet food and cardiovascular diseases, for food companies;
- Designing a poster to reduce school energy consumption;
- Surveying and raising awareness of parents' views about air pollution around the school, and measuring particulate pollution around the school.

Such short-term projects can use the framework for SSIBL and be aligned with curriculum requirements. As well as different time spans for SSIBL, they can also be structured by guidance from the teacher through to more open approaches. Structured inquiries are an important stage in helping make explicit the knowledge and procedures necessary to carry out more open inquiry to students, and to help them build up requisite knowledge and skills such as carrying out and analysing surveys, and importantly taking action based on evidence (Banchi and Bell 2008).

Scaffolding is also crucial in helping students turn authentic questions into inquiry activities. Drawing on social constructivist principles, Singer et al. (2000) have devised design principles including context leading to a 'meaningful defined problem space that provides intellectual challenges for the learner' (p. 167). A 'driving question' initiates the inquiry based on students' real world experiences and, importantly, helps to apply 'emerging scientific understandings'. But the problem in an authentic driving question based on characteristics in the SSIBL model is recognising the underpinning scientific principles. The driving question is anchored through sub-questions which link scientific knowledge to the main question. Hence a driving question such as 'How can we cut down our school's fuel bills?' (Levinson et al. 2017) (https://www.parrise.eu/wp-content/uploads/2018/04/parrise-en-rgb.pdf) can be anchored by student research addressing sub-questions such as 'Where are the cold 'spots' in winter'? 'Where are the sources of heat? 'How far does the temperature drop between the heat source and the cold spot?' The inquiry can be supplemented through student surveys about where and times of the day they feel cold. Science ideas can be reinforced by manipulating models which help students to test diverse components to reduce heat transfer.

Andrea's example and her open reflection indicate where scaffolding is needed – in helping students to highlight the problem and show how knowledge about sugary drinks can be used to solve it. The problems of using scientific knowledge are interconnected with gathering data because drawing on relevant knowledge presupposes how evidence is built up and vice versa. For example, students could be shown, perhaps simplified, correlative data between consumption of sugary drinks and health problems, to what extent they support causal links and what further questions could be asked. Julien and Barker's (2009) research demonstrate that students information retrieving skills show little interrogation of content, so need to be directed to this end.

Krstovic's (2014) work demonstrates that correlation studies can be introduced as apprenticeship activities in which young people both learn the need for such studies and how to interpret them. For example, in Andrea's case, the students needed to understand that studies of the effects of sugary drinks on health cannot ethically be attained through experiment, i.e., by exposing one sample of the population to sugary drinks over a long period of time while measuring against a control population. They could, for example, discuss different instances when experiments would be helpful, and when correlation studies would be more appropriate. When making inferences from relatively large sets of data, they would need to be aware of sources of bias, sample size, reliability and different types of variables. Before the students engage in SSIBL activities, they would need to have built up experience through other contexts in understanding the rationale for correlation studies. Otherwise, the cognitive hurdles are too onerous to overcome in one or two lessons. These are complex skills for both PSTs and students to grasp. Scaffolding surveys is crucial to effective gathering of relevant data. Caitlin's exemplary tasks supported the students in writing surveys but are unlikely to be effective unless the students have some time to build up expertise in writing short surveys and gathering data, possibly through interdisciplinary programmes such as ‘collapsed days’ (Harris and Ratcliffe 2005).

4.4 Implications

SSIBL activities do not lend themselves to one single disciplinary approach. What knowledge is needed depends on the question asked. Why young people smoke is a complex problem relying on psychological and identity constructs (e.g., youth culture), an understanding of the politics and economics behind advertising and distribution of tobacco products, legal regulations and broader social attitudes. Hence SSIBL needs to be containable by asking focused and feasible driving questions (Levinson et al. 2012; Singer et al. 2000). Adopting SSIBL could benefit from co-design frameworks (Kyza and Nicolaidou 2017; Kyza and Georgiou 2014) where cross-disciplinary teams of teachers with commonly identified purposes design resources and strategies for SSIBL approaches.

Martina’s demonstration of the smoking machine not only gave students the means to interrogate the way biological processes are represented but to identify the knowledge required for their surveys. Getting the ‘wrong answer’ is always a risk in open-ended inquiries but only if the main focus of the answer is conceptual knowledge (which is the case in the curriculum in England at present). In these cases, scientific knowledge is an input to promote deeper critical thinking rather than an output.

These case studies were carried out completely within the school context but can form stepping stones to activities outside the school, for example, with community action groups. Research by Dutta and Chandrasekharan (2017) on urban community farming in Mumbai emphasises the importance of practice in grounding knowledge to support further practice and action. The lesson here about SSIBL for teachers, science departments, schools, possibly in collaboration with community groups and informal science centres, is twofold. First that starting from action, the commitment to change is a means of exploiting and activating knowledge as well as generating interest and new questions, and, secondly, drawing on ‘teachable moments’, those opportunities presented by contemporary everyday events which can be linked to the curriculum.

Acknowledgements The research leading to these results is part of the PARRISE project, which has received funding from the European Union’s Seventh Framework Programme for research, technological development and demonstration under grant agreement no 612438. In particular, we would like to thank colleagues Christina and Katerina Ottander at Umea University, Sweden, for providing information on the Swedish science curriculum, and Michiel van Harskamp at Utrecht University for his support in designing the figures. Our thanks also go to all the pre-service and experienced science teachers who worked with us during the project, giving their time and valuable feedback on the SSIBL model.

References

Abd-El-Khalick, F., Boujaoude, S., Duschl, R., Lederman, N., Mamlok-Naaman, R., Hofstein, A., Niaz, M., Treagust, D., & Tuan, H.-L. (2004). Inquiry in science education: International perspectives. *Science Education, 88*(3), 397–419.

Banchi, H., & Bell, R. (2008). The many levels of inquiry. *Science and Children, 46*(2), 26.

Boerwinkel, D. J., Knippels, M. C., & Waarlo, A. J. (2011). Raising awareness of pre-symptomatic genetic testing. *Journal of Biological Education, 45*(4), 213–221.

Boerwinkel, D. J., Swierstra, T., & Waarlo, A. J. (2014). Reframing and articulating socio-scientific classroom discourses on genetic testing from an STS perspective. *Science & Education, 23*, 485–507.

Bouillon, L., & Gomez, L. (2001). Connecting school and community with science learning, real world problems and school/community partnerships as contextual scaffolds. *Journal of Research in Science Teaching, 38*(8), 878–898.

Brickhouse, N. (2011). Scientific literacy for bringing in the outsiders. In C. Linder, L. Ostman, D. Roberts, P.-O. Wickman, G. Ericksen, & A. MacKinnon (Eds.), *Exploring the landscape of scientific literacy* (pp. 193–204). New York: Routledge.

Buxton, C. (2006). Creating contextually authentic science in a "low-performing" urban elementary school. *Journal of Research in Science Teaching, 43*(7), 695–721.

Calabrese Barton, A., & O'Neill, T. (2008). Counter-storytelling in science: Authoring a place in the worlds of science and community. In R. Levinson, H. Nicholson, & S. Parry (Eds.), *Creative encounters: New conversations in science, education and the arts* (pp. 138–159). London: The Wellcome Trust.

Chinn, C., & Malhotra, B. (2002). Epistemologically authentic inquiry in schools: A theoretical framework for evaluating inquiry tasks. *Science Education, 86*, 175–218.

Davis, E., & Miyake, N. (2004). Explorations of scaffolding in complex classroom systems. *Journal of the Learning Sciences, 13*(3), 265–272.

Dawson, C. (2000). Selling snake oil: Must science educators continue to promise what they can't deliver? *Melbourne Studies in Education, 41*(2), 121–132.

Day, S., & Bryce, T. (2011). Does the discussion of socio-scientific issues require a paradigm shift in science teachers' thinking? *International Journal of Science Education, 33*(12), 1675–1702.

Dutta, D., & Chandrasekharan, S. (2017). Doing to being: Farming actions in a community coalesce into pro-environment motivations and values. *Environmental Education Research.* https://doi.org/10.1080/13504622.2017.1392485.

Ekborg, M., Ottander, C., Silfver, E., & Simon, S. (2013). Teachers' experience of working with socio-scientific issues: A large scale and in depth study. *Research in Science Education, 43*(2), 599–617.

Gormally, C., Brickman, P., Hallar, B., & Armstrong, N. (2009). Effects of inquiry-based learning on students' science literacy skills and confidence. *International Journal for the Scholarship of Teaching and Learning, 3*(2), 1–22, Article 16.

Gutierez, S. (2015). Integrating socio-scientific issues to enhance the bioethical decision-making skills of high school students. *International Education Studies, 8*(1), 142–151.

Hargreaves, A., & Shirley, D. (2009). The persistence of presentism. *Teachers College Record, 111*(11), 2505–2534.

Harris, R., & Ratcliffe, M. (2005). Socio-scientific issues and the quality of exploratory talk—What can be learned from schools involved in a 'collapsed day' project? *The Curriculum Journal, 16*(4), 439–453.

Hodson, D. (2003). Time for action: Science education for an alternative future. *International Journal of Science Education, 25*(6), 645–670.

Ingersoll, R., & Strong, M. (2011). The impact of induction and mentoring programs for beginning teachers: A critical review of the research. *Review of Educational Research, 81*(2), 201–233.

Julien, H., & Barker, S. (2009). How high-school students find and evaluate scientific information: A basis for information literacy skills development. *Library & Information Science Research, 31*(1), 12–17. https://doi.org/10.1016/j.lisr.2008.10.008.

Knippels, M. C., & van Dam, F. (2017). PARRISE, promoting attainment of responsible research and innovation in science education, FP7 – Rethinking science, rethinking education. *Impact, 5*, 52–54.

Koker, M. (1996). *Students' decisions about environmental issues and problems: An evaluation study of the Science Education for Public Understanding (SEPUP) Programme*. University of Southampton, PhD thesis. http://ethos.bl.uk/OrderDetails.do?uin=uk.bl.ethos.362538

Kolstø, S. (2001). Scientific literacy for citizenship: Tools for dealing with the science dimension of controversial socio-scientific issues. *Science Education, 85*(3), 291–310.

Krstovic, M. (2014). Preparing students for self-directed research-informed actions on socio-scientific issues. In L. Bencze & S. Alsop (Eds.), *Activist science and technology education* (pp. 399–418). Dordrecht: Springer.

Krstovic, M. (2017). Learning about youth engagement in research-informed and negotiated actions on socio-scientific issues. In L. Bencze (Ed.), *Science and technology education promoting wellbeing for individuals, societies and environments* (pp. 93–114). Dordrecht: Springer.

Kyza, E., & Georgiou, Y. (2014). Developing in-service science teachers' ownership of the PROFILES pedagogical framework through a technology supported participatory design approach to professional development. *Science Education International, 25*(2), 186–206.

Kyza, E., & Nicolaidou, I. (2017). Co-designing reform-based online inquiry learning environments as a situated approach to teachers' professional development. *CoDesign, 13*(4), 261–286.

Lederman, N., Antink, A., & Burton, S. (2014). Nature of science, scientific inquiry, and socio-scientific issues arising from genetics: A pathway to developing a scientifically literate citizenry. *Science & Education, 23*, 285–303.

Lee, H., Yoo, J., Choi, K., Kim, S., Krajcik, J., Herman, B., & Zeidler, D. (2013). Socio-scientific issues as a vehicle for promoting character and values for global citizens. *International Journal of Science Education, 35*(12), 2079–2113.

Levinson, R. (2009). The manufacture of aluminium and the rubbish-pickers of Rio: Building interlocking narratives. *School Science Review, 90*(333), 119–124.

Levinson, R., & the PARRISE Consortium. (2017). Socio-scientific inquiry-based learning: Taking off from STEPWISE. In L. Bencze (Ed.), *Science and technology education promoting wellbeing for individuals, societies, and environments* (pp. 477–502). Dordrecht: Springer.

Levinson, R., & Turner, S. (2001). *Valuable Lessons*. London: The Wellcome Trust.

Levinson, R., Hand, M., & Amos, R. (2012). What constitutes high quality discussion in science? Research from the Perspectives on Science course. *School Science Review, 93*(344), 114–120.

Levinson, R., Knippels, M.C., van Dam, F. et al. (2017). *Science and society in education. Socio-scientific inquiry-based learning: Connecting formal and informal science education with society*. https://www.parrise.eu/wp-content/uploads/2018/03/parrise-en-rgb.pdf

Millar, R. (2006). Twenty first century science: Insights from the design and implementation of a scientific literacy approach in school science. *International Journal of Science Education, 28*(13), 1499–1521.

Millar, R., & Osborne, J. (1998). *Beyond 2000: Science education for the future*. London: King's College.

Owen, R., MacNaghten, P., & Stilgoe, J. (2012). Responsible research and innovation: From science in society to science for society, with society. *Science and Public Policy, 39*, 751–760.

Ravetz, J. (2004). The post-normal science of precaution. *Futures, 36*(3), 347–357.

Roberts, D. (2011). Competing visions of scientific literacy: The influence of a science curriculum policy image. In C. Linder, L. Ostman, D. A. Roberts, P.-O. Wickman, G. Erickson, & A. MacKinnon (Eds.), *Exploring the landscape of scientific literacy* (pp. 11–27). New York: Routledge.

Rocard, M., et al. (2007). *Science education now: A renewed pedagogy for the future of Europe*. Brussels: European Commission.

Roehrig, G., & Luft, J. (2004). Constraints experienced by beginning secondary science teachers in implementing scientific inquiry lessons. *International Journal of Science Education, 26*(1), 3–24.

Roth, W.-M. (1997). From everyday science to science education: How science and technology studies inspired curriculum design and classroom research. *Science & Education, 6*(4), 373–396.

Roth, W.-M., & Lee, S. (2004). Science education as/for participation in the community. *Science Education, 88*(2), 263–291.

Ryder, J. (2002). School science education for citizenship: Strategies for teaching about the epistemology of science. *Journal of Curriculum Studies, 34*(6), 637–658.

Sadler, T. (2009). Situated learning in science education: Socio-scientific issues as contexts for practice. *Studies in Science Education, 45*(1), 1–42.

Schön, D. (1983). *The reflective practitioner. How professionals think in action*. London: Temple Smith.

Shatkin, J. (2013). *Nanotechnology: Health and environment risks*. Boca Raton: CRC Press.

Simonneaux, L. (2001). Role-play or debate to promote students' argumentation and justification on an issue in animal transgenesis. *International Journal of Science Education, 23*(9), 903–927.

Singer, J., Marx, R. W., Krajcik, J., & Clay Chambers, J. (2000). Constructing extended inquiry projects: Curriculum materials for science education reform. *Educational Psychologist, 35*(3), 165–178.

Stake, R. (1995). *The art of case study research*. London: Sage.

Sutcliffe, H. (2011). *A report on responsible research & innovation*. Brussels: European Commission.

Von Schomberg, R. (2014). The quest for the 'right' impacts of science and technology: A framework for responsible research and innovation. In J. van den Hoven et al. (Eds.), *Responsible innovation 1: Innovative solutions for global issues* (pp. 33–50). Dordrecht: Springer Science.

Voogt, J., Westbroek, H., Handelzalts, A., Walraven, A., McKenney, S., Pieters, J., & de Vries, B. (2011). Teacher learning in collaborative curriculum design. *Teaching and Teacher Education, 27*(8), 1235–1244.

Wierstra, R., & Wubbels, T. (1994). Student perception and appraisal of the learning environment: Core concepts in the evaluation of the PLON physics curriculum. *Studies in Educational Evaluation, 20*, 437–455.

Wilson, C., Taylor, J., Kowalski, S., & Carlson, J. (2010). The relative effects and equity of inquiry-based and commonplace science teaching on students' knowledge, reasoning, and argumentation. *Journal of Research in Science Teaching, 47*(3), 276–301.

Ruth Amos is a lecturer in science education at UCL Institute of Education, London. Prior to teaching for 13 years in London schools, Ruth was an environmental analytical research chemist. She works on the pre-service science teacher PGCE programme and the MA in Science Education. She has authored articles and book chapters on socio-scientific issues, initial teacher education and learning beyond the classroom. She was part of the EU PARRISE project and has contributed to teacher development programmes in China, India and the UAE. Her current research interests include learning through nature in primary and secondary schools.

Marie-Christine Knippels holds a PhD in Genetics Education and has been an Assistant Professor in Science Education at the Freudenthal Institute, Utrecht University, since 2007.She completed a postdoc research project on moral reasoning in genomics-related dilemmas. Her research interest is in SSI-based science education to foster scientific and democratic citizenship. She has authored articles on socio-scientific issues, opinion-forming, yo-yo thinking and systems thinking. She teaches in undergraduate and master science courses for future teachers and science education researchers. She has been involved in various European projects and networks, and led and coordinated the EU-FP7 PARRISE project from 2014 to 2018.

Ralph Levinson is Reader in Education at the Institute of Education University of London. After teaching science in London schools for 12 years he worked for the Open University where he wrote the first distance learning pre-service science teachers course. He has authored books and articles on socio-scientific issues, political aspects of science education, chemistry education, and arts in science education. He has been a keynote speaker at many national and international meetings, played a prominent role in European projects and has strong links to educational research communities in Brazil, South Korea, New Zealand, Australia, Canada, Occupied Territories (Palestine), as well as Europe. His present research interests focus on the role of neoliberalism in science education policy and school student engagement in scientific research.

Chapter 5
Critical and Active Public Engagement in Addressing Socioscientific Problems Through Science Teacher Education

John Lawrence Bencze, Sarah El Halwany, and Majd Zouda

5.1 Introduction

Scholars and others have long promoted school science that educates students about *socioscientific issues* (SSIs)—such as possible merits of pesticides, nuclear power and food additives—that seem relevant to decisions to which they may contribute as responsible citizens. Among arguments for this advice are that decision-makers can benefit from both *constitutive* (e.g., standards of practice like variable control) and *contextual* (e.g., ethical, ideological & cultural influences) values (e.g., Bingle and Gaskell 1994). Although school science systems have tended to prioritize education about constitutive values, they have generally struggled to attend to many contextual values. Particularly challenging has been attention to adverse effects on individuals, societies and environments of influences on fields of science and technology (and myriad other entities) of powerful pro-capitalist individuals and groups (e.g., financiers, banks, trade agreements & corporations) (e.g., Carter 2008). Highlighting such problematic relationships may, for example, cast fields of science and technology in a bad light and, therefore, discourage some students from choosing to pursue further related education and careers.

For about the last decade, some successes in educating students about capitalism-related harms linked to science and technology—along with possible personal and civic actions to address them—appear to have been achieved through implementation

J. L. Bencze (✉) · S. El Halwany · M. Zouda
OISE, University of Toronto, Toronto, ON, Canada
e-mail: larry.bencze@utoronto.ca; sarah.elhalwany@mail.utoronto.ca; majd.zouda@mail.utoronto.ca

M. Evagorou et al. (eds.), *Science Teacher Education for Responsible Citizenship*, Contemporary Trends and Issues in Science Education 52,
https://doi.org/10.1007/978-3-030-40229-7_5

of 'STEPWISE'-related perspectives and practices[1] (Bencze 2017). Adoption of this framework (elaborated below) across a breadth of educational situations seems, however, relatively limited. Accordingly, in this chapter, after a review of socioscientific issues education, we discuss our collaborative case study of three earlier efforts to facilitate science educators' implementation of STEPWISE. Results appear to suggest that critical and action-oriented science and technology education may be more relevant to school science systems and societies and environments if promoters prioritize development of supportive *dispositifs*; that is, assemblages of cooperating living, non-living and symbolic entities appropriate to each educational context.

5.2 Status of Socioscientific Issues Education

For about the last 50 years, scholars and others have urged school science systems to acknowledge studies of professional science and, accordingly, teach students about interrelationships among fields of science and technology and societies and environments (STSE) (Pedretti and Nazir 2011). Despite recommendations for such more relational education, however, school science systems tend to prioritize instruction in relatively easy-to-teach and assess 'products' (e.g., laws, theories & inventions) and skills (e.g., techniques for determining mass & volume) of science and technology—often dissociated, for example, from their societal contexts of production and uses (Levinson 2018).

Where school science systems have emphasized STSE relationships, *controversies* within them tend to be prioritized. Known in many places in the world as *socioscientific issues* (SSIs) education (Sadler 2011), students are often asked to debate conflicting positions of 'stakeholders'—including politicians, company executives, citizens, activists, scientists and others—regarding possible-harmful effects of products (e.g., biotechnologies) of science and technology on individuals, societies and environments. In such approaches, a major tendency—with exceptions—has been to invite students to interrogate (often in social situations) data and claims and then develop carefully-argued *personal* positions on issues (Levinson 2010; Zeidler 2014). Such argumentation-based approaches appear to provide students with significant benefits, including development of socioscientific reasoning skills (Sadler et al. 2007) and conceptions about the nature of science (Khishfe and Lederman 2006). Personal expertise like this can have considerable relevance for students, helping them make decisions about many personal and civic matters that may have connections with fields of science and technology.

[1] STEPWISE is the acronym for *Science & Technology Education Promoting Wellbeing for Individuals, Societies & Environments*. Some details about this framework are provided below, but it is a schema that organizes teaching/learning goals in ways that encourage students to engage in critical and active civic participation. For more information, refer to: www.stepwiser.ca

Although there is much to celebrate about socioscientific issues education, there also appear to be reasons for concern. Foci on controversies, rather than on harms, may be particularly limiting. Often associated with fields of science and technology are many threats to wellbeing of individuals, societies and environments—including, for example, from a plethora of products and services such as: household cleaning and hygiene products pesticides, nuclear power, tobacco and pharmaceuticals. Considered particularly worrisome are numerous harms to biotic and abiotic systems due to climate change said to be caused, in large part, by human burning of fossil fuels (Forzieri et al. 2017).

Given existence of many potential and realized harms linked to fields of science and technology, it may be tempting to blame practitioners, managers, etc. of such fields. Such blame may be, however, at least partly unwarranted. Instead, it is apparent that much responsibility for such harms can be attributed to powerful pro-capitalist individuals (e.g., financiers) and organizations (e.g., transnational corporations). Indeed, there is much evidence and argumentation to suggest that, based to a great extent on *neoliberal* socio-economic principles (e.g., Springer et al. 2016), powerful pro-capitalist entities have managed to orchestrate myriad living, nonliving and symbolic entities—including many governments, transnational economic groups (e.g., World Trade Organization), transportation and communication lines, think tanks (e.g., Atlas Network), banks, universities, transnational trade agreements, etc.—into a global capitalist *dispositif*[2] that appears to be promoting such ideals as: personal entrepreneurship and competitiveness, cost externalizations, 'public' service privatizations and business de-regulations.

Particularly important for functioning of the capitalist dispositif are fields of science and technology as instruments of commodity production, mobilization and disposal. Much research in science and technology studies suggests that capitalist associations with fields of science and technology often have compromised goals, methods, dissemination and uses of products in ways that can be associated with harms like those noted above (Mirowski 2011). School science systems, including through many recent 'STEM' education initiatives (Gough 2015), have not, however, apparently given adequate attention to such problematic associations (Carter 2008). Moreover, even when students conduct secondary research to delve into related socioscientific issues, access to information about adverse influences of pro-capitalist entities on science and technology may be significantly limited. Oreskes and Conway (2010), for instance, suggest that there have been many pro-capitalist efforts—such as through misleading science reports given to media—to create doubt in people's minds about evidence and arguments that would otherwise implicate private sector influences on production, distribution and disposal of commodities such as pesticides, cigarettes and nuclear power. Recent apparent tendencies of some politicians, corporations and think tanks, etc. to create confusion around science claims—such as regarding climate change research (Behr 2017)—may add to students' difficulties in adjudicating socioscientific controversies. Moreover, despite

[2] A dispositif (Foucault 2008) is a relatively organized network of living, nonliving and symbolic entities ('actants') that – like a machine – generally co-support each other in ways that serve common purposes.

emphases on students' logical decision-making, it is apparent that they are often placed in roles as *receivers* of knowledge from experts and/or people with power—creating a certain dependency on them. In analyses of socioscientific issues education, Levinson (2010), for example, concluded that most prioritized 'Deficit' (citizens needing to be informed) or 'Deliberative' (citizens engaged in discussions with fellow citizens) models of citizenship—both of which he claims place most citizens in deference to experts and/or people with power. Such learning outcomes seem relevant for participation in *representative* democracies, in which citizens only periodically influence governing—as they, for example, vote in elections every few years (Wood 1998). However, this arrangement can be problematic if leaders and/or experts do not advocate strong and deep attention to potential problems within controversies—which, as argued above, often appears to be the case in many democracies.

Based on evidence and arguments like that above that indicate problematic influences of pro-capitalist entities on governments and fields of science and technology, many scholars and others have urged science and technology educators (among others) to encourage and enable students to critically analyze knowledge production and dissemination systems and, where they identify harms, develop and implement action plans to address them (Hodson 2011). Regarding Levinson's (2010) categories of citizenship, students may, through 'Praxis' (e.g., self-led research), develop 'Dissent' (e.g., becoming critical of fields of science and technology and their relationships with members of societies and environments) and develop and implement 'Conflict' by, for example, engaging in socio-political actions to challenge power relations in ways that may lead to actors' conceptions of a better world. Students/citizens would, then, be involved in activities associated with more *participatory* forms of democracy—in which power is more widely-distributed across populations (Wood 1998).

5.3 Research Context and Methods

5.3.1 Research Context

Since September 2006, the first author here has been working with graduate students (including the second and third authors), teachers in primary, secondary, tertiary and after-school contexts and others to field-test the 'STEPWISE' science and technology education framework. This schema is designed to encourage and enable learners to use at least some of their education to develop and implement personal and social actions to address harms they perceive in relationships among fields of science and technology and societies and environments (STSE) (Bencze 2017). To motivate students to, in a sense, spend some of their educational 'wealth' (e.g., science literacy) on improving conditions around them, a key aspect of this framework is to encourage and enable students to *self*-direct primary (e.g., studies) and secondary (e.g., Internet searches) research to generate findings that may help motivate and

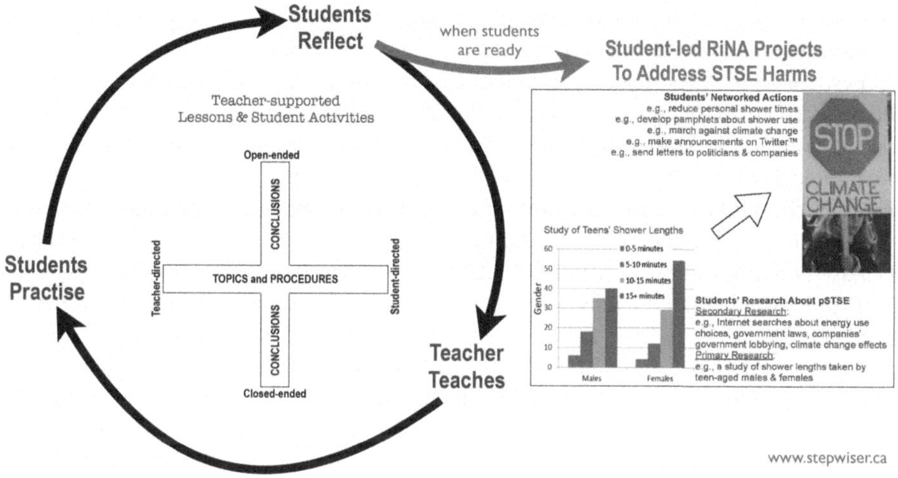

Fig. 5.1 STEPWISE pedagogical framework

direct possible actions they may negotiate with peers and others (e.g., parents, friends). As illustrated in Fig. 5.1, a typical such 'research-informed and negotiated action' (RiNA) project may involve secondary research to learn more about climate change, studies of peers' shower lengths (primary research) to learn about local contexts of energy uses and, based on findings (and, likely, knowledge from other sources), develop actions such as pamphlets for community members, letters to politicians and social media campaigns.

For many reasons, perhaps most likely due to school science systems' tendencies to over-emphasize teaching, learning and assessment of widely-accepted products and common skills of science and technology (Levinson 2018), students often struggle to self-direct often-unpredictable RiNA projects without some prior teacher mentoring. Among many possible approaches to help students to gain expertise, confidence and motivation to self-direct RiNA projects, our research suggests that the 3-phase constructivism-informed schema in Fig. 5.1 can be quite effective. With brief notes for each, the three phases involve: (i) Students Reflect. To prioritize students' personal contexts and agency, they are first asked to reflect on and *express* (e.g., in writing, as sketches or in conversation) their current attitudes, skills and knowledge ('ASK') related to common commodities (e.g., cell phones, fast food, fighter jets). Such activities would be mainly student-directed (SD) and open-ended (OE) to help ensure students feel comfortable expressing *their* ASK; (ii) Teacher Teaches. To ensure students' ASK are not limited by their varied (perhaps disadvantaged) backgrounds and their self-led inquiries (e.g., due to aforementioned problems of knowledge availability), teachers are urged to provide relatively teacher-directed (TD) and closed-ended (CE) instruction to share with students some important ASK about STSE relationships (e.g., role of private sector in influencing research and dissemination) and sample RiNA projects (including some by students of similar ages and stages of development). At the same time, to promote critical and deeper conceptions of such ASK, teachers also may urge students to analyse and evaluate them through activities (e.g., critique of a sample RiNA project) that may have some SD and open-endedness; and, (iii) Students Practise. To deepen students' expertise, confidence and motivation for SD/OE RiNA projects, they are asked to design and conduct small-scale RiNA projects to address harms they identify in STSE relationships—receiving teacher encouragement and support (i.e., some TD/CE aspects), as needed. At the end of a 3-phase cycle like that above, the teacher could ask students to return to the *Students Reflect* stage and, depending on the teacher's judgement of student progress, either engage students in another 3-phase pedagogical cycle (often in a different subject area) or ask students to conduct student-led (SD/OE) RiNA projects to address harms in STSE relationships of interest to them.

After about a decade of field-testing of STEPWISE-informed approaches in diverse educational contexts, much evidence suggests—supported by educational and other theoretical perspectives—that teaching/learning strategies developed based on schema like that in Fig. 5.1 have helped many students to eventually self-direct RiNA projects. Teachers have, for example, enthusiastically shared results in publications such as the STEPWISE edited book (Bencze 2017) and in two issues of the *Journal for Activist Science & Technology Education* (goo.gl/N00b3s; bit.ly/2JGIgtf) edited by science teachers. Regarding a central theme of this book, RiNA projects supported by pedagogical perspectives and practices as described in Fig. 5.1 (and elaborated in Bencze 2017) may provide considerable *relevance* through socioscientific issues education aimed at encouraging responsible citizenship at least in terms of the following two major aspects of the framework:

- *Student-centeredness*. A major priority of STEPWISE is to encourage and enable students to self-direct—and thereby *identify with*—RiNA projects to address problematic STSE relationships that seem *personally* relevant/important to them. This may be achieved, for instance, through students' self-reflections regarding merits of commodities about which they are likely to be familiar and, therefore, find relevant, and through RiNA projects in the *Students Practise* and *Student-led RiNA Projects* aspects of the schema in Fig. 5.1;
- *Holistic and problematizing instruction*. Although students can learn much from their inquiries, it is apparent that teachers need to more directly teach some often difficult-to-discover important aspects of STSE relationships and RiNA projects. For example, students can be taught to see commodities like cell phones as not just isolated (*punctualized*, in Callon's (1991) terms) entities but ('de-punctualized') as elements of complex global networks of living, nonliving and symbolic entities largely serving interests of relatively few pro-capitalist individuals and groups while creating various problems for individuals (e.g., poor labour conditions), societies (e.g., wealth disparities) and environments (e.g., species losses). In association with such more holistic and problematizing instruction, teachers can share with students cases of community-based actions that have attempted to address relevant harms. Their subsequent actions to address harms they perceive can have considerable relevance for wellbeing of individuals, societies and environments.

Although some educators have had significant successes using STEPWISE-informed perspectives and practices that may have aspects of relevance for students' responsible citizenship, implementation tends to be limited to relatively rare educational contexts in which a favourable *dispositif* exists and/or develops. Earlier studies suggested a teacher's successes promoting RiNA projects were aided by factors such as congruent official curricula, supportive administrators and colleagues and the teacher's compatible views about the nature of science (Bencze and Krstovic 2017). To perhaps confirm and more-deeply investigate needs for and characteristics of dispositifs that may help mobilize STEPWISE-informed perspectives and practices across more educational contexts, we conducted a collaborative study of three cases of our earlier efforts as researcher/facilitators of educators' action research cycles as they attempted to promote student-led RiNA projects to address possibly-problematic STSE relationships. Results of our collaborative case study may be relevant to educators and others wanting to promote such critical and action-oriented projects.

5.4 Research Methodology and Methods

To further investigate roles of dispositifs in mobilizing STEPWISE-informed perspectives and practices, we engaged in a *collaborative* case study (Schwandt and Gates 2018)) of our respective earlier studies (each led by one of us) of three science educators' uses of them to encourage and enable learners to implement self-directed research-informed and negotiated actions (RiNA) to address harms they perceived in relationships among fields of science and technology and societies and environments (STSE). With our interests in *mobilization*, we chose to review our previous studies of three educators who worked in very different educational contexts, including: (i) Nurul Hassan, teaching microbiological laboratory techniques to adults (ages 20–50) enrolled at a community college near Toronto, Ontario; (ii) Mirjan Krstovic, teaching science for students in grades 9–10 (ages 14–16) in a suburban public high school near Toronto, Ontario; and, (iii) Tomo Nishizawa, teaching science for students in grades 10–11 (ages 15–17) at a private international school in Venezuela. Our earlier research in the three cases combined rationalistic and naturalistic characteristics (Guba and Lincoln 2011)—the former in the sense that we planned to learn about factors such as teachers' views about the nature of science while, more naturalistically, we strived to elaborate unexpected findings. Data collected in these earlier studies included: lesson plan materials; student activity sheets; audio-recordings of semi-structured interviews with the teacher and a few (about five) volunteering students; anecdotal records taken during interviews; and, teachers' reflective journals. Data were analyzed using constructivism-informed constant comparative methods (Charmaz 2014) to generate themes and categories for learning outcomes and factors possibly contributing to them. In the course of our reviews of data and reports from these three cases, we each generated a written summary description (of about 1200 words) of each case and a 'dispositif map (Fig. 5.2, 5.4 and 5.5) to represent major living, nonliving and symbolic entities that appeared to cooperate to enable each educator to successfully implement RiNA projects. After each of us produced our separate case descriptions and accompanying dispositif maps, we then provided each other with critical and supportive feedback until we were all satisfied with our three summaries and maps. In doing so, we

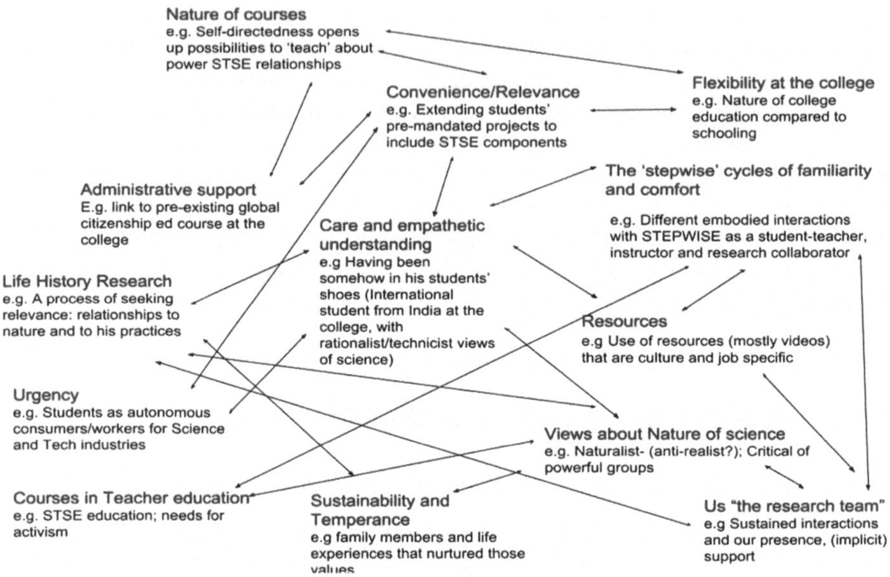

Fig. 5.2 Dispositif map of Nurul's successes promoting RiNA projects

then generated suggestions (in the Coda section, below) for actants that scholars and others may consider using in promotion of dispositif development in other educational contexts.

5.5 Summaries of Dispositifs for Mobilization

5.5.1 STEPWISE in a Technical College (Sarah El Halwany)

Nurul Mohammad is a contractual instructor who has been teaching in the Industrial and Microbiology Diploma Program for 4 years at a local community college in the Greater Toronto Area. Graduates from this program may work as biotechnicians (mostly in quality assurance) in food and pharmaceutical companies. Prior to his role as an instructor, Nurul had been an international student at the college—where he earned a technoscience diploma. He, thus, identifies with most of his students at the college because of such common experiences; but, also, because they, like him, 'come from India, are international students and they struggle with lack of family support, lack of money…'.

Nurul was introduced to STEPWISE while enrolled in a science teacher education course for which Larry (first author here) was the instructor, later enriching his understanding through masters and doctorate-level graduate courses conducted by Larry. Such educational experiences then led him to share STEPWISE perspectives and practices with his college-level students—choosing to do so in microbiology lab project courses, as well as in a 'physics for food science technology' course.

As a research team, our interactions with Nurul led to: (1) various action research projects (e.g., Schaffer et al. 2017) to help him facilitate STEPWISE pedagogy in his lab courses, (2) a reflective journal detailing his actions and students' reactions to STEPWISE as implemented in a lecture-based course 'physics for food science technology' and (3) research in which Nurul reflected on his commitments to STEPWISE practices in relation to his life history (El Halwany et al. 2017). We base our analysis of his dispositif-of-relevance—given in Fig. 5.2—on those three main formal sources of interactions and an emerging fourth source, resulting from our latest interactions with Nurul as part of my thesis fieldwork.

In our earliest efforts to facilitate action research in his context, Nurul found a relevant connection between STEPWISE and the college's mandatory global citizenship education course. As a result, he was able to gain administrative support on his practices with STEPWISE by alluding to a closely-aligned required course at the college. Nurul often mentioned how his students had never heard of the acronym STSE, mainly because it was specific to Ontario school science curricula and most of his students have graduated from international high schools that do not incorporate this component. Interestingly, absence of STSE in his students' former education was another main actant that motivated him to 'expose them to something that they have never heard about before' (Nurul's journal). He felt that the 'STSE component was very important and yet missing from an education that is too reduced to cause-effect relationships and technical aspects of science and technology.'

Nurul recognizes his somehow higher level of 'autonomy' in teaching at a college compared to being a school teacher, which allows him to 'tweak a little bit the areas of instruction time and the assessment criteria in order to introduce concepts of STSE.' Furthermore, he chose to implement STEPWISE in the microbiology project courses because they were mostly self-directed, giving him more room for teaching. As a matter of fact, and compared with ways he navigated implementation of STEPWISE in the 'physics for food science technology' (lecture-based) course, Nurul remarked that the microbiology labs are less structured and less heavy on 'delivering content' and 'preparing for exams,' which gave him more flexibility for implementation. In contrast, he struggled to accommodate correlational studies and research-informed actions in the lecture-based physics course, due to the large class size (90 students) and varying attendance of students, which 'didn't allow for sustained and deeper interactions.' Regarding microbiology project lab courses, students are expected to self-develop and implement protocols in microbiology with food and pharmaceutical applications (e.g., probiotics). Often, Nurul commented that the nature of those project courses facilitated teaching with a social and environmental lens, as they were already grounded in products of science and technology that students are likely to often purchase. For instance, when a group of students sought to 'evaluate efficiency of probiotic microorganisms against various antibiotics,' Nurul incited them to look at the 'bigger picture of science and technology' when designing their investigative protocols. As a result, students incorporated a second component to their inquiry that included a 'comparative study of natural and commercial probiotic products,' in which they surveyed peers' knowledge of presence of probiotic organisms in natural products and critically reflected on consumers' tendencies to buy commercial probiotics based on media and advertisements. This extension to their original experimental procedure 'didn't feel too much like extra work' (Nurul's comment), as it flowed naturally from their original protocols.

Allowing students to understand the 'bigger picture of science and technology' emanated from Nurul's own changing views on nature of science. After taking courses at OISE as part of his teacher education programme, he mentioned how he learned 'to see science education through a completely different lens. I learned to see the interconnected nature of all disciplines, I developed Naturalist views about knowledge production' (Nurul's reflections). His views about the nature of science were found to be closely-aligned with his views about what exists (his ontological views), as well as his axiological views about nature and sustainability. The life history research was an opportunity for Nurul to seek relevance into his commitments to STEPWISE by referring to earlier experiences with nature; his relations to family members who 'inculcated values of sustainability, temperance and harmony with nature' and who helped him developed a 'holistic view of the world.' As a result, he came to recognize importance of 'connecting science to other disciplines.'

Nurul's dispositif of relevance seemed greatly influenced by his *embodied* experiences with STEPWISE. Nurul experienced the framework as a student (in Larry's courses), an action-researcher (implementing it in his work context) and as a collaborator (participating and discussing research claims and ideas). Moreover, his ability to closely-associate his own former experiences with his students' experiences at the college allowed him to develop empathetic understanding and genuine care for an education that 'shifts from the familiar and well-beaten track of memorizing content knowledge and regurgitating on the exam into the realms of social, environmental and ethical issues.'

Because of close cultural associations with his students, Nurul often chose to use examples for points-of-discussion on STSE relationships that have cultural significance—including: 'challenges of e-waste in India'; and, 'water contamination in rural India.' He also referred to and problematized media in India, such as 'fairness cream' advertisements, in which a potential suitor is humorously portrayed as being deceived by white skin of an Indian girl who had applied a fairness cream. Those examples created opportunities for students to self-identify and critically problematize taken-for-granted assumptions about race and skin colour in relation to consumption of S&T products (cosmetics) in their home country. Moreover, Nurul encouraged students to think about finding solutions to STSE problems that are relevant to their degree in biotechnology. For instance, he showed an Indian scientist who appeared on TED talks presenting his invention of edible cutlery as a way to mitigate use of plastic ones. Also, recognizing how milk is an important part of people's diet in India, Nurul showed them an example of scientists using milk to make edible food wrappers. Another video further related to biotechnicians' work with microorganisms, showed how two young scientists succeeded in breaking down plastics with bacteria. All those examples were purposefully shown to further sensitize students to become more involved in STSE issues, allowing them to somehow 'see themselves' (whether culturally or professionally speaking) as agents of change regarding such issues.

5.5.2 *STEPWISE in a Suburban Public High School (John Lawrence Bencze)*

After about 5 years of relative frustration in encouraging science educators and others to help students to self-direct RiNA projects to address harms they determined in STSE relationships, Mirjan Krstovic, whom I only knew as one of several students enrolled in my online graduate course about history, philosophy and sociology of science, asked me to help him explore different approaches for enabling students to develop more 'authentic' conceptions of the nature of science ['NoS']. At our first face-to-face meeting in Sept., 2011, Mirjan agreed that his teaching about NoS could be explored through his efforts to encourage and enable students to conduct student-led RiNA projects. It was my contention that more 'realistic' conceptions of NoS would include considerations of STSE relationships, particularly regarding associations among capitalists and scientists and engineers.

To achieve our goals, I worked with Mirjan as a researcher/facilitator for three successive school years (six 5-month semesters, two per year) to encourage him to reflect on his pedagogical perspectives and practices, develop new ones, implement revised perspectives and practices in his teaching, repeating such action research cycles as we proceeded. Largely framing my pedagogical suggestions was the 'STEPWISE' schema in Fig. 5.1. Throughout this work, besides data collected as described above (for all three cases), Mirjan completed some theoretical schema, such as his perspectives on NoS in terms of the *Scientific Theory Profile*[3] (Loving 1991) and his priorities for science teaching and learning regarding his completed repertory grid[4] (Gaines and Shaw 1993).

By the end of our 3-year collaboration, Mirjan's abilities to encourage and enable students to self-direct RiNA projects progressed to the point that we agreed that most students' projects were quite *sophisticated*. Figure 5.3 summarizes one such project, organized in terms of a conception of RiNA projects I developed based on Roth's (2001) framework for science-technology relationships. In this schema,

[3]The STP consists of a 2-dimensional grid. Its horizontal axis spans a continuum ranging from Rationalist through Naturalist positions regarding the nature of theory negotiation. Rationalists tend to believe in highly systematic methods of science, including rational judgements about theory. Naturalists, by contrast, assume that conduct of science is highly situational and idiosyncratic, depending on various factors, including psychological, social, cultural and political influences. The vertical axis depicts a continuum reflecting the truth value of knowledge, with Realist through Antirealist positions. Realists tend to believe that science knowledge can correspond to reality, while (extreme) Antirealists claim that each person's constructions are valid. More moderate Antirealists believe in useful knowledge.

[4]A repertory grid is a two-dimensional rectangle suggesting, in our case, associations between teaching strategies and learning outcomes.

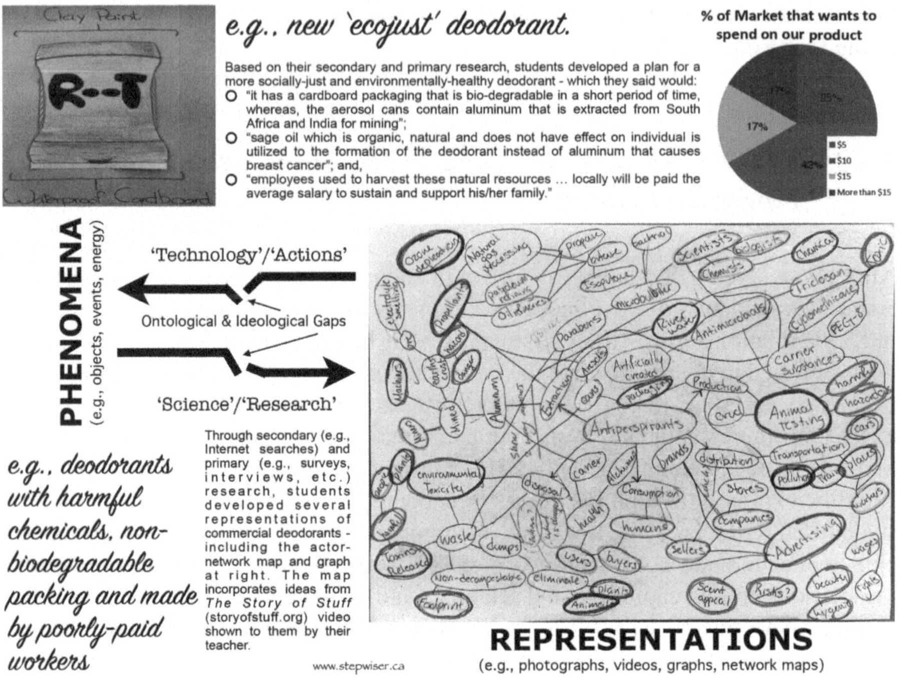

Fig. 5.3 A well-developed RiNA project

translations from Phenomena ➔ Representations denotes 'research' (and 'science) and translations from Representations ➔ Phenomena relate to 'actions' ('technology'). Both kinds of translations may, however, be 'ineffective'—given that differences in the nature of entities may lead to *ontological* gaps in translations (e.g., a picture not fully representing air pollution) and, perhaps more importantly, people may purposely misrepresent (*ideological* gaps) phenomena (e.g., climate change) (Bencze and Carter 2015). Students did seem to, likely tacitly, come to acknowledge such mis-translations in the context of uses of aspects of actor-network theory (ANT) (Latour 2005). As suggested in Fig. 5.3, for instance, their analyses of commercial antiperspirants (Phenomena ➔ Representations), depicted as an actor-network map of them, seemed to align with suggestions (e.g., Pierce 2013) that commodities should not be seen as isolated entities but, rather, as part of larger networks of living, nonliving and symbolic entities. Meanwhile, in the opposite direction (Representations ➔ Phenomena), ingredients (e.g., organic sage oil) of their new deodorant was to align with their claims about it.

Until Mirjan approached me, relatively few teachers seemed willing and able to encourage and promote critical and action-oriented projects to address harms they perceive in STSE relationships. In his case, however, it was apparent that—as shown in Fig. 5.4—a highly cooperative set of actants appeared to develop to enable projects like that depicted in Fig. 5.3. Although it is philosophically invalid from an ANT perspective to discuss individual actants, some major (likely-interrelated) fac-

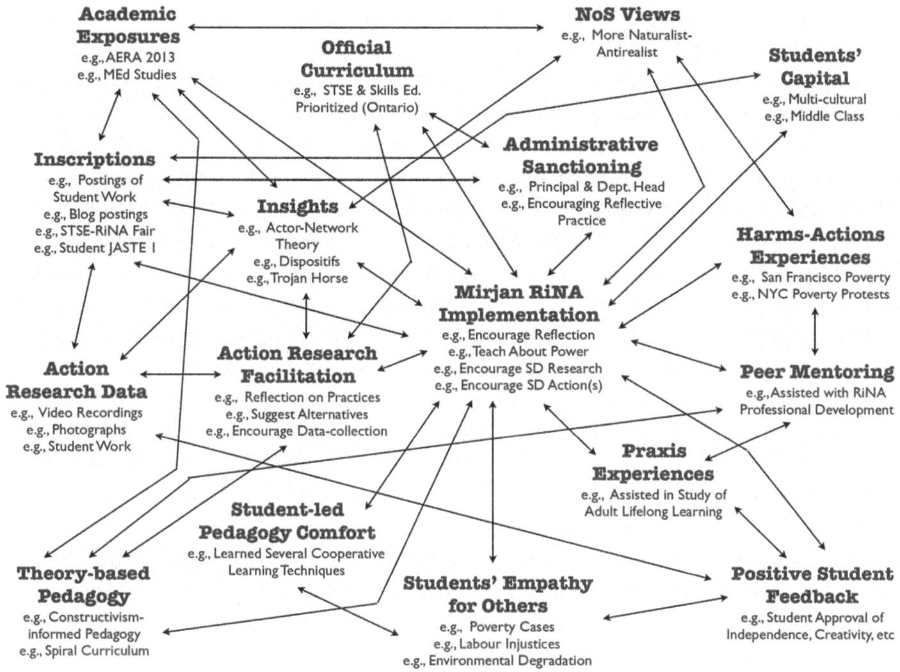

Fig. 5.4 Dispositif map of Mirjan's successes promoting RiNA projects

tors contributing to Mirjan's successes were apparent to us. To begin with, Mirjan often mentioned (as also noted by several other teachers) that promotion of RiNA projects to address perceived harms in STSE relationships would be much less likely if STSE and 'Skills' (e.g., for experimentation) education were not the first two overall goals of the local science curriculum. Indeed, it seemed especially helpful that STSE education was first listed as the first of three overall goals in the latest version of the curriculum (MoE 2008).

Despite prominence of STSE education (especially) in the local official curriculum, it appeared to us that many teachers struggled to implement this educational component (Pedretti and Nazir 2011). To a great extent, therefore, it seems logical (borne out by our studies) that Mirjan possessed and developed personal and professional characteristics conducive to RiNA project promotion. He seemed to be a particularly energetic teacher, highly motivated and confident in executing student-led activities and action research—which appeared to exist prior to my work with him, mainly through previous action research in which he was involved. But, a factor that seemed to steadily develop as I worked with him was his views about the nature of science—which, by our third year of collaboration—came to relatively-strongly support more *Naturalist-Antirealist* positions on Loving's (1991) *Scientific Theory Profile*. Earlier research suggested that teachers holding such views were more likely to facilitate student-led activities (perhaps to supplement more teacher-led lessons) (Bencze et al. 2006), a tendency that is necessary for RiNA projects.

Teachers holding particular pedagogical perspectives and talents may be somewhat limited in opportunities to implement them, depending on various material, intellectual, psychological and other resources. In Mirjan's case, many of these were in place, but it appeared to be particularly enabling to work in schools that had administrative personnel (e.g., principal and department head) who—although not, necessarily, supportive of particular aspects of RiNA project implementation—encouraged ongoing teacher critical reflective practice (praxis). With curricular and contextual supports in place, along with his relevant personal and professional qualities, Mirjan seemed very amenable to learn about (often from me) and/or develop pedagogical perspectives and practices that appear aligned with encouraging and enabling youth to develop and implement effective RiNA projects to address harms they perceived linked to influences of powerful entities on fields of science and technology. His effective uses of ANT, for instance, seemed especially important—including, as evidenced in Fig. 5.3, in teaching about ANT via uses of *The Story of Stuff* (storyofstuff.org) videos and the Trojan horse metaphor (i.e., commodities portrayed in overly-positive ways, while harbouring potential personal, social and/or environmental harms). His gradual encouragement of infrequently-used correlational studies in science education (Bencze 1996) also seemed quite helpful. Finally, that most students in schools in which Mirjan worked came from families with moderate-to-high socio-economic levels suggests they may have had access to cultural and social capital (Bourdieu 1986) conducive to creative and effective RiNA projects. But, Mirjan also noted that students in his eleventh-grade chemistry class (like the ones whose project is highlighted in Fig. 5.3) who had been in his tenth-grade science class were particularly comfortable developing self-led RiNA projects as compared to students learning about these for the first time.

5.5.3 STEPWISE in an International Private High School (Majd Zouda)

Tomo Nishizawa is a Japanese teacher who has international teaching and learning experiences, including completing her bachelor's degree in sciences and later her master's degree in educational studies in Canadian universities. During our collaboration with Tomo, she worked as a high school biology teacher in an International Baccalaureate (IB) programme implemented by a school in Venezuela. Our work with Tomo began when she was looking on the Internet for pedagogical/instructional frameworks that appear to contextualize science (hence, making it more 'relevant' for students). After coming across the STEPWISE framework, Tomo contacted Larry (i.e., first author) and requested using his pedagogical framework in her grade 10 science and grade 11 biology classes. We collaborated with Tomo over two academic years, focusing mainly on grade 11. We conducted regular online meetings with her, in which we brainstormed and discussed plans and implementations. Tomo also shared with us her written reflections. We examined students' experiences implementing STEPWISE through minor and major assignments, and by conducting Skype™ interviews with some of these students.

Our analyses suggest that there were interrelating personal and contextual factors that, along with some aspects of the STEPWISE framework, affected how relevant it was (perceived) for/by students and teachers (Fig. 5.5). By personal factors, we mean elements that affected individuals' motivation and commitment. In the case of Tomo, her science-related educational goals and her beliefs about nature of science apparently allowed her to see STEPWISE as a pertinent tool to achieve these goals. For example, Tomo holds relatively Naturalist (refer above) perspectives about NoS. Although she believes that 'knowledge generation is systematic in the sciences,' she also strongly stated that scientists' work and ways of thinking are influenced by personal and social contexts. These views seemed to have affected Tomo's goals of teaching/learning science to include supporting 'students to have a

Fig. 5.5 Dispositif map of Tomo's successes promoting RiNA projects

contextualized understanding of how science operates in society' and allowing spaces for examining STSE problems in her science classroom.

Although Tomo's views and beliefs about NoS allowed her to perceive STEPWISE as a relevant pedagogical framework, some contextual factors encouraged her to adopt and apply it in her classrooms. These include some aspects of the IB curriculum and supportive administrators and colleagues. In 2016, the IB organization started their first assessment of its NoS component (International Baccalaureate Organization 2014). Tomo thought that some aspects of IB-NoS, such as recognizing 'the human face of science' (ibid, p. 10), well-resonated with STEPWISE principles. Similarly, Tomo perceived possible connections between STEPWISE and the IB curriculum through its 'IB learner profile' which, for example, aims for learners to be principled, caring, reflective … etc. and in the aims set specifically for science curricula (e.g., 'become critically aware, as global citizens, of the ethical implications of using science and technology' (p. 18)). Although IB aims and views on NoS may not fully align with STEPWISE goals and perspectives, Tomo saw in these components enough space (and legitimacy) to implement STEPWISE. She also managed to make adaptations to fit STEPWISE into the school curriculum. On the other hand, the IB curriculum also posed some challenges; the exam-driven curriculum and time limits exerted some pressure over Tomo's implementation—especially when considering students' limited expertise with some of the STEPWISE 'adds-on' components (i.e., teaching about STSE issues through actor-network analysis) and needs to scaffold them through its three pedagogical stages. This 'collection' of perhaps hindering factors may have interfered with how relevant STEPWISE was perceived.

Difficulties Tomo may have perceived seemed to have been mitigated, however, through continuous support provided by the school's administration and by her colleagues. For example, the principal and the superintendent were encouraging of including research-based practice and practice-based research in their school—with strong foci on teachers exploring new practices to apply in the school. At the same time, some teacher colleagues were generous in supporting Tomo through collegial favours to implement STEPWISE pedagogy and research. For example, one colleague offered his office to confidentially Skype-meet with some of Tomo's students and volunteered to organize these meetings, collect consent forms, and ensure smooth progress of the research. Together, these supportive co-workers seem to have created an atmosphere of comfort and confidence for Tomo to implement the framework and for students to get involved in it.

In terms of the students, there were also some other factors that affected the level of their engagement in and commitment to their research and actions. Among these factors, are how close an STSE issue was to students' day-to-day life (e.g., the oil pollution of the main lake in the city), urgency of the issue (Zouda et al. 2017) and the length and depth of engagement in their research-action projects (Zouda et al. 2016). At times, the heated socio-political situation in Venezuela provided a rich and stimulating context for students to engage in urgent issues, such as scarcity of basic goods and roles of science and technology in addressing this problem. However, at other times, the low security in this tense situation prevented some students from reaching out to the public to implement their preferred actions (although some enthusiastic students took the risk to do so). In one academic semester, this unsettled situation also prevented students from regularly attending classes, interrupting their engagement in their projects. Nevertheless, Tomo managed to 'tweak' these projects—reframing them in ways directly related to urgency of students' lives. For example, she asked students to imagine that they are politicians with strong backgrounds in science and/or technology, and to use such expertise to suggest solutions that might improve the socio-political situation in Venezuela. Such project-framing increased students' investment in their work and made it more relevant for them.

Additionally, when focusing on students' choices of, investments in and commitment to actions, not only urgency of their socio-political context seemed to have shaped these, but also their social class, the graded nature of the STEPWISE school project and students' expertise and reflective practices. As mostly members of the Venezuelan upper-class, students tended to see prosperity and social wellbeing through vantage points of their class and took 'relevant' actions that supported it (Zouda et al. 2017). Some students took strategic actions (e.g., Facebook™ pages, educating the public and younger students) as a result of reflective practices regarding their possible effectiveness. However, with time limits and pressure of school work, and because taking actions was a component of a graded school-project, other students relied on familiar and easy-to-construct (in terms of their expertise) forms of actions to complete their projects, with little thorough thinking about how effective might be these actions. For example, Facebook™ pages were sometimes created as a checklist or as a 'generational trend,' rather than an informed, well-studied form of action. In her reflections, Tomo argues for needs to *deliberately* coach students in their reflective practices and teach them about what forms of actions might best serve their causes; hence, rendering them relevant to their issues, purposes, contexts and expertise.

In summary, in this case, a dynamic combination of different factors seemed to have supported relevant adoption and implementation of STEPWISE as a pedagogical framework. These include Tomo's goals and beliefs of NoS, spaces available in the curriculum, supportive administration and co-workers, and the socio-political contexts of the school and the city. Additionally, how the project is framed, students' backgrounds and expertise, their length and depth of involvement in the projects, and the level of scaffolding tended to affect the level of students' engagements in the projects and their actions.

5.6 CODA

Research conducted prior to the study reported here had suggested that teachers' abilities to encourage and enable students to self-direct research-informed and negotiated action (RiNA) projects to address problematic relationships among fields of science and technology and societies and environments (STSE) largely depends on existence of a supportive *dispositif*; that is, a set of cooperating living, nonliving and symbolic actants (Bencze and Krstovic 2017). It was previously concluded, for instance, that, although a teacher may hold views about the nature of science congruent with promotion of RiNA projects to address power-related problems in STSE relationships, they may be limited to do so without—among numerous factors—significant 'official sanctioning,' such as aligned curriculum goals and administrative (e.g., principal) and collegial (e.g., other science teachers) supports.

In light of findings like those above concerning importance of dispositifs for facilitation of teachers' efforts to enable students to self-direct RiNA projects, we sought to explore the extent to which it could be used to further understand implementation of RiNA projects in diverse educational contexts. More specifically, we conducted a collaborative case study of our respective earlier efforts to facilitate teachers' action research to learn more about RiNA project implementation in three very different situations—in terms, for instance, of age groups, geographic locations and social class. In concert with writing summaries of our respective findings from these three cases, we constructed (for the first time in this work) *dispositif maps*; that is, possible relationships among sets of actants that seem greatly-relevant to promotion of student-led RiNA projects.

Through this collaborative case study, among new findings were that, besides alignment of government curricula, 'official sanctioning' also may be provided through existence of congruent institutional mission statements (as with the college studied here). It also became apparent that, in addition to importance of certain views about science, teachers' adherence, for instance, to congruent values positions—such as those inherent to recent *ethics-of-care* movements (Blum and Murray 2017)—seemed helpful. Arguably, such perspectives may lead teachers to share (i.e., through more direct instruction) such perhaps difficult-to-discover attitudes, skills and knowledge as those regarding adverse effects of influences of powerful people and groups on fields of science and technology and actions people have taken to address some of them. Such education, in turn, may lead students to accommodate more collectivist views of existence and, perhaps accordingly, more societal responsibilities. In terms of the central theme of this book, which is to (broadly) highlight for science teacher education relevance of roles of socio-scientific issues in helping students to become responsible citizens, this chapter provides support for two broad claims; that is, i) encouraging and enabling student-led RiNA projects appears highly relevant in light of myriad personal, social and environmental harms linked to powerful individuals and groups; and, ii) those working to mobilize student-led RiNA projects across multiple educational contexts may benefit from emphases on relevant dispositif development.

Although it may be tempting to use claims like those provided above to work to assemble a dispositif that encourages and enables learners to self-direct RiNA projects that address power-related harms that interest them, it seems clear that one cannot fully predict or manage dispositif formation. Using Deleuze and Guattari's (1987) *rhizome* metaphor, for example, knowledge systems develop in unpredictable and uncertain ways. Such a situation was, indeed, evident in distinctiveness of the three dispositifs in this study. On the other hand, given the extent to which dispositif maps overlapped (e.g., implementation often benefiting from congruent views about science), scholars and others wanting to mobilize STEPWISE-informed perspectives and practices across multiple contexts may find useful several elements they find common to the three dispositifs (and corresponding summaries) provided here.

Acknowledgements True to our adherence to actor-network theory and the dispositif concept, content of this chapter should be seen as an amalgamation of myriad living, nonliving and symbolic actants. Symbolically, for instance, this work is driven by various senses of injustice. Meanwhile, the extent to which this chapter deals with them seems connected to multiple technologies, animate and inanimate. Integrated into all of these, however, we are extremely grateful to long-term commitments to this project by the three teachers highlighted in this study (Nurul Hassan, Mirjan Krstovic & Tomo Nishizawa) and numerous others over about the last decade and knowledge-generation assistance from many graduate students who have been part of this project at various times.

References

Behr, H. (2017). The populist obstruction of reality: Analysis and response. *Global Affairs, 3*(1), 73–80.

Bencze, J. L. (1996). Correlational studies in school science: Breaking the science-experiment-certainty connection. *School Science Review, 78*(282), 95–101.

Bencze, J. L. (Ed.). (2017). *Science & technology education promoting wellbeing for individuals, societies & environments*. Dordrecht: Springer.

Bencze, J. L., & Carter, L. (2015). Capitalists' profitable virtual worlds: Roles for science & technology education. In P. P. Trifonas (Ed.), *International handbook of semiotics, vol. 1 & 2* (pp. 1197–1212). Dordrecht: Springer.

Bencze, L., & Krstovic, M. (2017). Resisting the Borg: Science teaching for common wellbeing. In J. L. Bencze (Ed.), *Science & technology education promoting wellbeing for individuals, societies & environments* (pp. 227–276). Dordrecht: Springer.

Bencze, L., Bowen, M., & Alsop, S. (2006). Teachers' tendencies to promote student-led science projects: Associations with their views about science. *Science Education, 90*(3), 400–419.

Bingle, W. H., & Gaskell, P. J. (1994). Scientific literacy for decisionmaking and the social construction of scientific knowledge. *Science Education, 78*(2), 185–201.

Blum, A., & Murray, S. J. (2017). *The ethics of care: Moral knowledge, communication, and the art of caregiving*. New York: Routledge/Taylor & Francis.

Bourdieu, P. (1986). The forms of capital. In J. G. Richardson (Ed.), *The handbook of theory: Research for the sociology of education* (pp. 241–258). New York: Greenwood Press.

Callon, M. (1991). Techno-economic networks and irreversibility. In J. Law (Ed.), *A sociology of monsters: Essays on power, technology and domination* (pp. 132–161). London: Routledge.

Carter, L. (2008). Globalisation and science education: The implications for science in the new economy. *Journal of Research in Science Teaching, 45*(5), 617–633.

Charmaz, K. (2014). *Constructing grounded theory* (2nd ed.). New York: Sage.

Deleuze, G., & Guattari, F. (1987). *A thousand plateaus: Capitalism and schizophrenia* (B. Massumi, Trans.). Minneapolis: University of Minnesota Press.

El Halwany, S., Bencze, L., Hassan, N., Schaffer, K., Milanovic, M., & Zouda, M. (2017). *Exploring connections between a college instructor's relationships to nature and to his practices on socio-scientific issues: A life history approach.* Paper presented at the semi-annual conference of the European Science Education Research Association, Dublin, Ireland.

Forzieri, G., Cescatti, A., Silva, F. P., & Feyen, L. (2017). Increasing risk over time of weather-related hazards to the European population: A data-driven prognostic study. *The Lancet Planetary Health, 1*(5), e200–e208.

Foucault, M. (2008). *The birth of biopolitics: Lectures at the Collége de France, 1978–1979* (M. Senellart, Ed.). New York: Palgrave Macmillan.

Gaines, B. R., & Shaw, M. L. G. (1993). Knowledge acquisition tools based on personal construct psychology. *Knowledge Engineering Review, 8*(1), 49–85.

Gough, A. (2015). STEM policy and science education: Scientistic curriculum and sociopolitical silences. *Cultural Studies of Science Education, 10*(2), 445–458.

Guba, E. G., & Lincoln, Y. S. (2011). Competing paradigms in qualitative research. In N. K. Denzin & Y. S. Lincoln (Eds.), *The SAGE handbook of qualitative research* (pp. 105–117). Thousand Oaks: Sage.

Hodson, D. (2011). *Looking to the future: Building a curriculum for social activism.* Rotterdam: Sense.

International Baccalaureate Organization. (2014). *Diploma programme: Biology guide.* Retrieved from http://www.sdgj.com/ckfinder/userfiles/files/9a6ff27578e18e3f1023a4dc046beef1.pdf

Khishfe, R., & Lederman, N. G. (2006). Teaching nature of science within a controversial topic: Integrated versus nonintegrated. *Journal of Research in Science Teaching, 43*(4), 395–418.

Latour, B. (2005). *Reassembling the social: An introduction to actor-network-theory.* Oxford: Oxford University Press.

Levinson, R. (2010). Science education and democratic participation: An uneasy congruence? *Studies in Science Education, 46*(1), 69–119.

Levinson, R. (2018). Realizing the school science curriculum. *The Curriculum Journal, 29*(4), 522–537. https://doi.org/10.1080/09585176.2018.1504314.

Loving, C. C. (1991). The scientific theory profile: A philosophy of science model for science teachers. *Journal of Research in Science Teaching, 28*(9), 823–838.

Ministry of Education [MoE]. (2008). *The Ontario curriculum, grades 9 and 10: Science.* Toronto: Queen's Printer for Ontario.

Mirowski, P. (2011). *Science-mart: Privatizing American science.* Cambridge, MA: Harvard University Press.

Oreskes, N., & Conway, E. (2010). *Merchants of doubt.* London: Bloomsbury Press.

Pedretti, E., & Nazir, J. (2011). Currents in STSE education: Mapping a complex field, 40 years on. *Science Education, 95*(4), 601–626.

Pierce, C. (2013). *Education in the age of biocapitalism: Optimizing educational life for a flat world.* New York: Palgrave Macmillan.

Roth, W.-M. (2001). Learning science through technological design. *Journal of Research in Science Teaching, 38*(7), 768–790.

Sadler, T. (Ed.). (2011). *Socio-scientific issues in the classroom: Teaching, learning and trends.* Dordrecht: Springer.

Sadler, T. D., Barab, S. A., & Scott, B. (2007). What do students gain by engaging in socioscientific inquiry? *Research in Science Education, 37*(4), 371–391.

Schaffer, K., Milanovic, M., El Halwany, S., Hassan, N., Zouda, M., & Bencze, L. (2017). *Inertial tensions in promoting socio-political actions among future technoscience technicians.*

Paper presented at the semi-annual conference of the European Science Education Research Association, Aug. 21–25, Dublin, Ireland.
Schwandt, T., & Gates, E. F. (2018). Case study methodology. In N. K. Denzin & Y. S. Lincoln (Eds.), *The Sage handbook of qualitative research* (5th ed., pp. 341–358). Los Angeles: Sage.
Springer, S., Birch, K., & MacLeavy, J. (Eds.). (2016). *The handbook of neoliberalism*. New York: Routledge.
Wood, G. H. (1998). Democracy and the curriculum. In L. E. Beyer & M. W. Apple (Eds.), *The curriculum: Problems, politics and possibilities* (pp. 177–198). Albany: SUNY Press.
Zeidler, D. L. (2014). Socioscientific issues as a curriculum emphasis: Theory, research and practice. In N. G. Lederman & S. K. Abell (Eds.), *Handbook of research on science education* (Vol. II, pp. 697–726). New York: Routledge.
Zouda, M., Nishizawa, T., & Bencze, L. (2016). *What do youth know about socioscientific issues? Reflecting for socio-political actions*. A paper presented at the annual conference of American Educational Research Association, Washington, DC, April, 2016.
Zouda, M., Nishizawa, T., & Bencze, L. (2017). In the eye of the hurricane': Using STEPWISE to address urgent socio-political issues in Venezuela. In J. L. Bencze (Ed.), *Science & technology education promoting wellbeing for individuals, societies & environments* (pp. 339–357). Dordrecht: Springer.

John Lawrence Bencze, (PhD, MSc, BSc, BEd) is an Associate Professor (Emeritus) in Science Education at the University of Toronto (1998-present). Prior to this role, he worked for 15 years as a science teacher and as a science education consultant in Ontario, Canada. His research programme emphasizes critical analyses—drawing on history, philosophy, sociology, etc.—of science and technology, along with student-led research-informed and negotiated socio-political actions to address personal, social and environmental harms associated with fields of science and technology. Recent publications include two edited books about proactive citizenship. He also is co-editor of an open-source activist journal (goo.gl/ir7YRj).

Sarah El Halwany is a PhD candidate at the Ontario Institute for Studies in Education (OISE), University of Toronto. She finished her bachelor of science and her master's in science education from the American University of Beirut, Lebanon. Her current thesis examines how affect produces and interrupts STEM subjectivities amongst international college students. Her research interests include STSE education, culturally relevant teaching and learning, and emotions in science education.

Majd Zouda is a Ph.D. candidate in Science Education at the Ontario Institute for Studies in Education, University of Toronto. Her doctoral dissertation focuses on STEM programs in independent schools in Canada. Majd holds a B.Sc. in Microbiology and M.Sc. in Medical Microbiology. Prior to pursuing a Ph.D. degree, she worked as a high-school science teacher and the head of the junior science department in an international school in Damascus, Syria. For her doctoral research, she has received the SSHRC doctoral fellowship (Social Sciences and Humanities Research Council), and the Ontario Graduate Scholarship (OGS). Majd has been actively involved in publications regarding socioscientific and STSE issues, STEM education and student activism.

Chapter 6
Supporting Teachers in the Design and Enactment of Socio-Scientific Issue-Based Teaching in the USA

Patricia J. Friedrichsen, Troy D. Sadler, and Laura Zangori

6.1 Introduction

The authors lead an on-going, five-year project working with teachers to co-design socio-scientific issues-based (SSI) curricula. In this chapter, we begin by describing our motivation for the project, followed by our perspective on collaborative professional development (PD). Next we describe our theoretical framework and our commitment to specific aspects of SSI teaching. The rest of the chapter consists of three design cases, presented in chronological order, representing how our thinking evolved over time. We close the chapter by reviewing what we have learned and then describe next steps.

Initially, the desire to collaborate with teachers and with each other was the motivation for our project. Collaboration is often defined as individuals working together to achieve a common goal (Katz and Martin 1997). Our research team's common goal is to support teachers and students in learning science through the use of scientific practices embedded within SSIs. In this project we collaborated with teachers to co-design SSI curricula, focusing on the development of tools and strategies to support SSI teaching and learning. In a collaboration, the members have complementary domains of expertise (John-Steiner et al. 1998). In our research team, Troy brings SSI expertise, Pat has expertise in teacher learning, and Laura has scientific modeling expertise. Our collaboration began in the autumn of 2013, shortly after the release of the *Next Generation Science Standards* (*NGSS*) (*NGSS* Lead States 2013) in the United States. Although the state of Missouri did not adopt the standards until

P. J. Friedrichsen (✉) · L. Zangori
University of Missouri, Columbia, MO, USA
e-mail: FriedrichsenP@missouri.edu; ZangoriL@missouri.edu

T. D. Sadler
University of North Carolina at Chapel Hill, Chapel Hill, NC, USA
e-mail: tsadler@unc.edu

M. Evagorou et al. (eds.), *Science Teacher Education for Responsible Citizenship*, Contemporary Trends and Issues in Science Education 52,
https://doi.org/10.1007/978-3-030-40229-7_6

4 years later, the local school district adopted *NGSS* as soon as they were released. The release of the new standards provided us with a window of opportunity to collaborate with teachers as they revised their curricula and challenged us to deepen our own understanding of how to implement the new standards using an SSI approach.

As we initially negotiated this collaborative research space, our individual research backgrounds and experiences sparked additional motivations for our collaboration. In Troy's previous work, his research team designed SSI curricula and researched student learning as classroom teachers implemented the units. Troy was further motivated to collaborate on this project because of the opportunity to co-design SSI curricula with an innovative high school biology teacher with curriculum design expertise. In Pat's previous research, using an etic approach, she studied science teachers' pedagogical content knowledge development which was often idiosyncratic depending on teachers' goals. She was further motivated to be part of this team because of her interest in collaborating with teachers who shared a common research-based vision for teaching (i.e., SSI). Laura's prior research focused on student learning and teacher thinking within modeling contexts in elementary classrooms. She found that teachers and students faced many challenges with modeling. She was looking for ways to help teachers see the need to overcome these challenges and incorporate modeling into their instruction. Laura's additional motivation to work on the project was that an SSI approach brought a level of complexity that necessitated scientific modeling as a sense-making tool. Our common goal of supporting teachers in using scientific practices embedded within an SSI context provided a context that allowed us to collaborate with each other and with teachers, while supporting our individual research interests.

6.2 Collaborative Professional Development Model

From the beginning of this project, we used a collaborative professional development (PD) model that emphasizes 'nurturing learning communities within which teachers try new ideas, reflect on outcomes, and co-construct knowledge about teaching and learning in the context of authentic activity' (Butler et al. 2004, p. 436). In our collaborative PD model, teachers collaborate with each other and with researchers, resulting in the sharing of both formalized and practical knowledge (Butler et al. 2004). In considering authentic teacher activity, we prioritize a) the importance of curriculum coherence (Schmidt et al. 2005) and b) teachers as curriculum designers. As researchers, our goal is to empower teachers to design, implement, and reflect upon an SSI-based approach to teaching science. To accomplish this goal, we co-design curricular materials with teachers, and develop tools and strategies to support the design and implementation of this approach. As we design PD, we draw upon the National Academies of Science, Engineering, and Medicine's (2015) recommendations for effective science PD: (1) active participation of teachers engaging in examples of effective instruction and analysis of student work, (2) content focused,

(3) alignment with district policies and practices, and (4) sufficient duration for implementation and reflection. As teachers implement their newly designed SSI curriculum, we support their reflection-on-practice (Schön 1983; Valli 1997) through follow-up interviews. In this project, we drew upon the aspects of collaborative PD described in this section. Over time, our ideas about collaborative PD evolved and these ideas are represented in the Design Cases section of this chapter.

6.2.1 Theoretical Framework

SSIs are contentious, societal issues with conceptual or procedural connections to science (Zeidler 2014). Working to understand and resolve these issues can (and, we argue, should) be informed by scientific evidence, but science alone cannot render solutions (Sadler and Zeidler 2005a, b). Take, for instance, climate change, arguably the most important SSI of our time. Scientists identified the problem, can monitor the problem, and can make predictions about what will happen given different responses to the problem. However, working toward climate change solutions requires political and economic actions, and social change in addition to scientific inquiry.

We take the view that being a responsible citizen requires that one engage in the problem-solving/solution-seeking associated with SSIs. Dealing with SSIs, like all wicked problems, is notoriously complex and requires a wide range of knowledge and competencies. We posit that these knowledge bases and competencies can be learned, but in order to be useful, knowledge and competencies require applications in context. Further, employing knowledge and competencies in complex contexts requires practice (Sadler 2009). SSI teaching represents a pedagogical approach to supporting learner development of knowledge, competencies, and abilities to apply these to understanding and problem-solving in the context of science-related societal issues. In our approach, we engage learners (or support teachers as they engage learners) directly in the exploration of SSIs (Friedrichsen et al. 2016). As a part of this exploration, students learn about the science content and practices necessary for understanding the issue. They also have opportunities to explore some of the social dimensions of the issue, which may include political, economic, or ethical factors, depending on this issue.

In this chapter, we report on a set of three design cases conducted iteratively within a broader design-based research project (Brown 1992; McKenney and Reeves 2013). An important dimension of the broader agenda is to generate an empirically supported framework for SSI teaching. This framework evolved through the course of the studies presented. When the work began (i.e., Design Case 1 described in the next section), our framework highlighted several commitments: (1) Instruction should begin with the SSI and subsequent learning experiences should continually connect back to the issue; (2) Students should have opportunities to engage with scientific practices, content, and evidence as they make sense of the issue; (3) Students should use current media to access information about issues, and they likely need support for building media literacy skills; (4) Students should have opportunities to explore the social dimensions of SSI; and, (5) Issue-based learning experiences should conclude with a culminating exercise so that learners can synthesize their understandings, competencies, and personal positions on the issue. Based on our student-centered approach to teaching, our framework purposefully focuses on the nature of student engagement. As the work has progressed, details regarding these aspects of SSI teaching have evolved and become better substantiated, but these basic commitments remain. The following Design Cases section of the chapter presents highlights from the studies we conducted based on these initial commitments and explicates some of the ways in which our ideas have changed.

6.3 Design Cases

6.3.1 Design Case 1: Collaborating with an Exemplary Biology Teacher

Research Focus Our initial focus was on developing a collaboration among the research team members and a local high school biology teacher. We sought to develop our understandings of the newly released *NGSS* by co-designing *NGSS*-aligned curriculum units situated within an SSI context. As we co-designed two curriculum units, our research focus was on student learning of science content and practices within the SSI context. For example, in the Climate Change unit, our research questions were: (1) In what ways did students' model-based explanations of carbon cycling, climate change, and the interrelationships between them change over time in response to a model-oriented SSI unit? and (2) What are the ways that students come to understand carbon cycling, climate change, and the interrelationships? (Zangori et al. 2017).

PD Design In the fall of 2014, we invited a local teacher, Kerri Graham, to collaborate with us. Pat had previously supervised student teachers in Kerri's classroom. Kerri was in the process of re-designing the Honors Biology curriculum to align with *NGSS* and welcomed our help with the re-design process. (Honors Biology is a course offered to 16-year-old students in the 10th grade and has a focus on college preparation with more challenging content). We met approximately every 2 weeks for 6 months during Kerri's planning period at her school. For our first unit, we chose a second-semester topic, evolution, giving us time to develop our collaboration and design the new unit. After considering the affordances and limitations of different SSIs, we selected antibiotic resistance (ABR) as the issue, since many of Kerri's students were interested in healthcare-related careers. We chose to focus on one *NGSS* practice, modeling, because we sought to better understand how to support students with this particular practice. After making these decisions, we invited a microbiologist to join our team. Collaborating with a scientist became a curriculum design feature that we often incorporated as we designed additional curriculum units.

We discussed the components of our SSI framework and developed a general outline for the unit. Next, different team members took the lead in developing components of the ABR unit. We used a video case to introduce ABR and designed a media-based lesson to engage students in exploring the issue in more depth. We developed an ABR lab, several modeling activities, and additional instructional resources to help students make sense of the ABR phenomenon (Williams et al. 2018). We also designed an activity for students to explore various stakeholders' perspectives (e.g., doctors, parents, pharmaceutical companies). For the culminating activity, each student developed a policy recommendation (at a local, state, federal or international level) to reduce the spread of antibiotic resistant bacteria. (For more detail of the design process, see Friedrichsen et al. 2016). During the implementation of the ABR unit, we continued to support Kerri as we collected student data.

In subsequent years, we continued our collaboration with Kerri to design a second SSI unit using the issue of climate change to teach ecology (Kinslow et al. 2017). For this unit, we invited an ecologist to join our design team. The students were introduced to the issue by taking a field trip to a local prairie to observe the effects of climate change. Due to a change in precipitation patterns, woody plants were beginning to replace the prairie grasses. Throughout the rest of the unit, students applied ecological concepts to the local prairie context. We continued to develop modeling tools and strategies to support students in making sense of climate change on both local and global scales. For example, we developed a modeling packet format which included the following features: an initial model and written explanation in response to a scenario, multiple revised models and written explanations, critique of previous models, and a final model and written explanation in response to the original scenario. For the culminating activity, students were asked to apply their knowledge by exploring the effects of climate change on a species indigenous to a different ecosystem.

Research To understand student learning related to science content and modeling practices, we collected student artifacts, including pre- and post-assessments, and their modeling packets. We also interviewed a sub-set of the students about their models. In the climate change unit, modeling supported students' understandings of causal mechanisms for transfer and transformation of carbon which were needed to make connections between carbon cycling and climate change (Zangori et al. 2017). In the ABR unit, we found students gained greater understanding of generalized natural selection, but had difficulty in understanding how natural selection occurred in bacteria (Peel et al. 2019).

Implications We learned that SSI teaching with an emphasis on NGSS-aligned practices, such as modeling, could produce desired student learning outcomes. We also learned how to better support students' modeling practices through the development of a modeling packet format. However, Design Case 1 had several limitations: results were based on a single teacher's enactment, our team provided extensive planning support to the teacher, and her honors classes were populated with high achieving students.

6.3.2 Design Case 2: Secondary Teachers Co-Designing Curriculum

Research Focus Based on the limitations of Design Case 1, we sought to explore processes for helping a broader range of teachers use SSI teaching. We wanted to find out how teachers, working in varied settings with more diverse students, might take up SSI teaching. Given the goal to work with a larger number of teachers, the kinds of individualized collaboration we employed in Design Case I would be impossible. We needed to adopt a PD model that could reach more participants, and we felt that it would be helpful to formalize some of the design rationales and principles used in (and for some, modified after) the first design case. We chose to concretize dimensions necessary for SSI planning and teaching through the development of several tools. These tools and how teachers used them became another focal point for our research.

PD Design In order to recruit participants, we sent invitations to teachers in a broad geographic area across our state. We purposefully invited teachers from schools of different sizes and community types (rural, suburban, and urban). We also targeted teachers with different levels of experience, from second year novices to 25-year veterans. In our communications with potential participants, we called attention to the fact that the opportunity would focus on using issues as contexts for teaching science and prioritize connections to the NGSS, particularly in terms of scientific practices. Ultimately, we worked with 19 high school biology, chemistry, and environmental science teachers. Many of these teachers participated as the sole representative from their school—for some, they were the only teacher for a particular course, but others worked as a part of a professional learning community (PLC) within their home institutions. Two teachers taught together in the same school, and five participants taught together and worked collaboratively in the same district.

We designed the PD as a series of workshops and working sessions with the goal of small teams of teachers co-designing SSI units that they could enact in their classes. The sessions began with an orientation meeting at a science teacher conference in the autumn. The group met for two full days the following March, and then again for 3 days in June when the teachers were on summer break. We used an online teacher networking platform to encourage continued communications and resource sharing between the face-to-face sessions. The sessions were designed such that the teachers could: (1) experience dimensions of SSI teaching as learners; (2) reflect on the pedagogy of those experiences; (3) access examples (e.g., SSI curricula and learning activities), samples (e.g., student work and rubrics), and relevant tools; and, (4) work with colleagues to design an SSI unit to meet the needs of their own classrooms. For more detail about this PD, see Peel et al. 2018.

The tools shared included a framework for SSI teaching, a heuristic for planning SSI units, and a guide for assisting in the selection of educationally productive issues. We fully describe the *SSI Teaching Framework* elsewhere (Sadler et al. 2017). In short, this Framework offers a pathway for sequencing SSI instruction with attention to student learning objectives. The pathway begins with student exploration of the issue followed by student engagement in scientific practices as they make sense of the underlying science content. Student development of science ideas and practices is complemented by opportunities to build socio-scientific reasoning skills (Romine et al. 2017). The sequence concludes with an issue-focused culminating activity that allows students to synthesize their ideas, practices, and reasoning. The *Planning Heuristic* provided a list of nine recommended steps for the successful design and development of SSI units. The *Issue Selection Guide* presented a list of ordered questions designed to encourage critical analysis of possible SSIs in terms of how effective they might serve as contexts for science teaching and learning. The three tools can be accessed on the project website (http://ri2.missouri.edu/content/Planning-Tools).

Research In order to explore how a diverse group of teachers engage in SSI design and teaching, as well as use tools that formalize SSI design principles, we collected data from multiple sources. These data included field notes taken during the workshops, interviews with teacher design teams during the workshops, individual teacher interviews following unit enactment, and the teachers' design products (i.e., their SSI curricular materials). Qualitative analyses have yielded numerous insights regarding the research foci.

As the teachers participated in the PD, they moved through two emergent, sequential phases (Hancock et al. 2019). In the first phase, each design team created a safe and shared space by identifying common topics taught, discussing tensions and discontentment related to their existing curricula, and exploring how the PD could be used to generate opportunities to address their tensions and discontentment. Once the design team established a safe and shared space, they explored potential SSIs for their curriculum unit. The design team's selected issue was based upon three considerations: an individual's passion for a particular issue, the ability to leverage existing resources, and their perceptions of the relevance of a given issue for their particular students.

All of the teachers were able to successfully participate in the development of SSI units, although many initially lacked curriculum design skills. The resulting design products varied extensively, and many did not incorporate all of the teaching elements called for in the *SSI Teaching Framework* although several did. Most of the participants implemented at least some dimensions of their planned units and when doing so, featured opportunities for students to explore the focal issue at the outset of instruction. However, several teachers reported that they struggled to keep the issue connected to the science content and practices throughout the unit. In addition, only half of the teachers who implemented were able to complete a culminating activity with their students. They cited time constraints and concerns about how to assess the culminating projects, many of which took the form of essays, organized debates, or student-generated artifacts such as posters.

In the PD, we encouraged teachers to focus on any of the NGSS practices they perceived fitted best with their units and the needs of their students. We provided explicit support for teacher use of the practices of modeling, argumentation, and computational thinking as evidence suggests that these are some of the more challenging practices for teachers to enact with learners. The extent to which scientific practices were meaningfully incorporated in the teachers' unit designs and enactments varied, and teachers' prior experiences with NGSS was a strong predictor of how well practices were incorporated. The ways in which participants experienced the PD relative to connections with colleagues from their schools served as another mediator for successful enactment. In general, participants with school colleagues at the PD were most successful. Participants who were the sole instructors for a course at their home institutions were also reasonably successful. However, participants who worked as a part of a PLC in their schools but were lone representatives at the PD tended to struggle with enactment, because they had the added challenge of bringing their units back to colleagues who had not been involved in the PD.

Results regarding teacher use of the planning tools varied. The *SSI Teaching Framework* was perceived by teachers as a useful tool particularly as they considered issues of lesson sequencing. However, the design products suggest that they struggled to make sense of how to situate socio-scientific reasoning, a key dimension of the *Framework*, with science content and practices. In contrast, the teachers indicated that the *Planning Heuristic* was not very useful to them, and there is little evidence to suggest that this tool was even used by many of them. Some of the teachers lacked experience in designing curricula which may have contributed to their perceptions of the usefulness of the *Planning Heuristic*. Finally, most teachers used the *Issue Selection Guide*, but they found it helpful for assessing issue choices they had already made and potential adjustments to the framing of an issue, instead of a tool for generating potential issues, as we had originally intended.

Implications These results offer several implications for further research and development. First, we intend to pay much closer attention to the natural groups that teachers work in. Rather than recruiting individual teachers, it seems prudent to consider teachers nested within PLCs. Working with PLCs wherein individual teachers have access to natural, school-based support mechanisms may help overcome some of the challenges associated with SSI teaching. Second, the results forced us to question our decision to leave open the selection of scientific practices. The findings suggest that a more focused approach may have ultimately been more supportive for teachers, particularly those less familiar with the NGSS. Third, tools such as the *SSI Teaching Framework* and the *Issue Selection Guide* can be helpful for teachers engaged in SSI planning and enactment. We need to adjust aspects of the *Framework* to account for aspects with which teachers struggled (viz., socio-scientific reasoning), but we will continue using these resources. We also intend to use the *Planning Heuristic* again; however, how the heuristic is featured in the PD and supports for using this tool need to be considered. Finally, an important implication of this phase is that while a culminating activity can be an effective exercise for student synthesis of ideas and practices (see Design Case I), it represents one of the more challenging dimensions of our model of SSI teaching. Our team needs to develop more effective ways to support teacher incorporation of this dimension of SSI pedagogy.

6.3.3 Design Case 3: Implementing SSI Teaching in an Elementary School

Research Focus The impetus to try SSI-based instruction within the elementary classroom was two-fold. First, we sought to explore how elementary teachers implement and grapple with the multiple dimensions inherent in issues-based learning as this is missing from the literature base. We also wanted to examine how elementary students learn within an SSI context, as this is emerging area of study (e.g., Evagorou 2011). However, since SSI-based instruction is naturally interdisciplinary, elementary classrooms seemed an ideal setting, as elementary teachers are generalists, teaching across disciplines within their daily work. Second, we wanted to explore how an SSI context could support teachers in understanding the purpose and utility of model-based teaching and learning. The few studies about elementary teacher knowledge and implementation of model-based instruction highlight the challenges they have in understanding the purpose and utility of modeling (Vo et al. 2015; Justi and Gilbert 2002). We theorized that situating modeling within an SSI would provide a utility and purpose to scientific modeling, since teachers could see their students develop and use scientific models to understand phenomenon, and they could apply their new scientific understanding when negotiating complex societal problems.

PD Design Based on results from our previous work (Design Case 2), we wanted to engage with an intact PLC as opposed to teachers whom we might recruit from across multiple schools. Therefore, we approached a third-grade PLC of four teachers from a local school to co-design and implement an SSI-based ecosystem curriculum. Because of what we had learned from Design Case 2 about the difficulties of supporting teacher curricular design for multiple scientific practices, we intentionally focused on the practice of modeling. The project team worked with the teachers to choose the focal issue of decreasing Monarch butterfly (*Danaus plexippus*) migration (an issue that was playing out within our state context), and whether or not we should conserve and/or restore prairies to return Monarch numbers to previous levels. The biological content focus of the unit was ecosystem interdependence.

Because we wanted the SSI curriculum to focus on the practice of modeling, the project team wrote the SSI and modeling-focused lessons prior to the workshop, and then co-designed the ecosystem lessons with the PLC during the workshop. Working with the PLC, we integrated modeling throughout the unit, wherein students used the practice to answer the question, ‘How do organisms interact within an ecosystem?’ The students were also asked to use their scientific models to provide evidence to answer the issue-based restoration question, ‘Should our school turn one soccer field into a garden to attract butterflies?’ In their final lesson, students responded to the following scenario:

> An elementary school has a prairie habitat in their backyard next to a soccer field. But more students want to be able to play soccer and have asked for another soccer field. The principal is thinking about turning the prairie habitat into a soccer field. Write a letter to the school principal about whether or not the prairie should be turned into a soccer field.

The PLC attended a four-day workshop, spending the first 2 days immersed in a SSI unit for high school students to provide the teachers with experiences in both modeling and SSI. The last 2 days of the workshop were spent introducing the teachers to the ecosystem curriculum, during which we worked with the PLC to design, modify, and adapt the lessons for their classrooms (full curriculum available at: http://ri2.missouri.edu/ri2modules/MONARCH/intro). We followed all four teachers through their implementation of the curriculum.

Research Our research focused on how the teachers perceived the individual SSI and modeling aspects of the curriculum, and how they valued this pedagogical approach. Our results indicated that the elementary teachers perceived SSI as a way to promote social responsibility and make real world issues relevant to their students. In addition, the teachers wrapped their understanding of modeling within the context of the SSI. The two seemed inseparable to the teachers as they focused their modeling reflections on how the students used their models to make sense of ecosystem interdependence and articulate their position on prairie restoration or conversation. As one of the teachers stated about her students using her model to articulate her position on the issue:

> I was so blown away with [student] who is a [below grade level] in reading and what she could share with me about her model was so above that level of reading. That was incredible to see that the kids that do have trouble with reading or expressing themselves can still have the same level of thinking with the model.

Implications Implications from this work suggest that elementary teachers are able to successfully navigate the challenges of discussing grade-level appropriate complex issues with social dimensions in their classrooms, and they also find SSI-based instruction as a productive way to contextualize science for their students. The teachers expressed how the cross-curricular nature of SSI felt familiar, because teaching across subject-domains occurs daily in the elementary classroom. This study also provided two important strategy implications for our team. First, the curriculum helped the teachers see how their third-grade students (8–9 years old) used models of their own design to make sense of ecosystem interactions and apply their understanding to the Monarch issue. Second, this work highlighted the importance of working with PLCs from idea conception through unit implementation. All four PLC members attended the workshop, worked with us to co-design the unit, focused weekly PLC meetings on lesson planning for the SSI unit, and taught the unit during the same time frame. These actions served as important support mechanisms for their individual enactments. Since the teachers were integral in co-designing the unit, the PLC had full buy-in enacting the curriculum. As they planned together and enacted the curriculum, the teachers reminded each other why they made certain choices during the unit design, and they checked with each other on whether modifications and adaptations were successful or not during the enactments.

6.3.4 Conclusions and Next Steps

Reflecting on the questions that drove each design case, we come to the following conclusions. In Design Case 1, we sought to understand the following questions: (1) Could this be a productive collaboration? (2) What do we learn in co-designing *NGSS*-aligned curriculum using an SSI-based approach? (3) How do we support teachers and students in the practice of modeling? We found that it was, indeed, a productive (and enjoyable) collaboration that we plan to continue. We greatly valued our collaboration with the microbiologist and plan to continue collaborating with scientists whose expertise aligns with a particular SSI. We learned students do learn *NGSS*-aligned science content using an SSI-based approach to curriculum design. We also learned how to support modeling through tools embedded in a modeling packet that included student prompts for initial and final models explaining a scenario, written explanations of their models, and critique and revisions of models.

Design Case 2 was driven primarily by the question: How do we support a larger, more diverse group of teachers in co-designing SSI curricula? To explore this question we designed a PD for 19 teachers from across our state; the teachers varied in experience and teaching context, ranging from rural to urban schools. For this PD, we designed a set of tools to support teachers in their curriculum design work, including the *SSI Teaching Framework*, the *SSI Planning Heuristic*, and the *Issue Selection Guide*. We learned that the teachers perceived the *SSI Teaching Framework* as useful, giving them an overview of our teaching approach. In general, most of the teachers struggled with curriculum design and they did not use the *Planning Heuristic*. Finally, rather than starting with the *Issue Selection* Guide, teachers chose issues based upon their own passion for a particular issue, existing resources, and their perceptions of student relevance. After selecting an issue based on a combination of these three considerations, teachers tended to use the *Issue Selection Guide* tool to confirm the appropriateness of their selected SSI. We also learned that it was more productive for us to support teachers in using a single *NGSS* practice, modeling, rather than having teachers choose from among the eight practices. From their implementation of the SSI units, we learned that teachers were challenged in keeping the content and issue connected throughout the unit, and teachers often omitted the final culminating project due to time constraints and assessment concerns. We also learned that teachers were more likely to implement their SSI unit and feel successful when they had the support of their teaching colleagues.

For Design Case 3, based on the lessons learned in the previous design case, we worked with one elementary PLC of four elementary teachers. Our questions were: (1) How will elementary students respond to an SSI curriculum unit? and (2) How will elementary teachers integrate modeling and SSI? We learned that elementary students were able to grapple with age-appropriate, ill-structured societal problems and use their scientific knowledge to take and justify a position on the issue. We learned that elementary teachers saw SSI-based instruction as a way to integrate social studies and science, and they saw it as a productive way to contextualize science for their students. We also learned that the teachers came to understand the purpose of modeling as they saw students using their models to make sense of the SSI.

We plan to continue our project with an ongoing focus on tools and strategies to support teachers and students as they engage in learning science within SSI contexts. We are currently developing new tools to address the challenges identified in the design cases described in this chapter. In particular, we are developing tools to aid teachers in keeping the science content and issue connected throughout their SSI unit, and we are developing shorter alternatives to the culminating activity used in Design Case 1 – ones that requires less time and are less writing intensive. In the summer of 2018, we began working on a new design case, collaborating with a secondary biology PLC comprised of six teachers who teach in a high-needs school. We collaborated to revise an existing SSI curriculum unit (from Design Case 2) and co-design a second SSI unit. This new design case is driven by a new set of questions: (1) Is our SSI approach viable with students from a wider range of backgrounds, interests, and skills? (2) As teachers implement their SSI units, what teacher-designed tools and strategies are developed?

Acknowledgements We would like to acknowledge the critical contributions of Kerri Graham, our collaborator who teaches high school biology. Without Kerri's involvement, this project would not have moved from the drawing board to the classroom. As researchers, we have learned a tremendous amount from collaborating with Kerri. We would also like to thank the many teachers and graduate students who worked on this project; in particular, we would like to acknowledge the contributions of the following graduate students: Jaimie Foulk, Tamara Hancock, Eric Hayes, Andrew Kinslow, Hai Nguyen, and Amanda Peel.

This work was funded, in part, by the National Science Foundation under Award Number IIA-1355406. Any opinions, findings, and conclusions or recommendations expressed in this material are those of the authors and do not necessarily reflect the views of the National Science Foundation.

References

Brown, A. L. (1992). Design experiments: Theoretical and methodological challenges in creating complex interventions in classroom settings. *The Journal of the Learning Sciences, 2*(2), 141–178.

Butler, D. L., Lauscher, H. N., Jarvis-Selinger, S., & Beckingham, B. (2004). Collaboration and self-regulation in teachers' professional development. *Teaching and Teacher Education, 20*(5), 435–455.

Evagorou, M. (2011). Discussing a socioscientific issue in a primary school classroom: The case of using a technology-supported environment in formal and nonformal settings. In *Socio- scientific issues in the classroom* (pp. 133–159). Dordrecht: Springer.

Friedrichsen, P., Sadler, T., Graham, K., & Brown, P. (2016). Design of a socio-scientific issue curriculum unit: Antibiotic resistance, natural selection and modeling. *International Journal of Designs for Learning, 7*(1). Available at: https://scholarworks.iu.edu/journals/index.php/ijdl.

Hancock, T., Friedrichsen, P., Kinslow, A., & Sadler, T. (2019). Forming design teams and selecting socio-scientific issues: A grounded theory study of how science teachers collaboratively design SSI-based curricula. *Science & Education, 28*(6–7), 639–667.

John-Steiner, V., Weber, R., & Minnis, M. (1998). The challenge of studying collaboration. *American Educational Research Journal, 35*(4), 773–783. Retrieved from http://www.jstor.org/stable/1163466.

Justi, R., & Gilbert, J. (2002). Science teachers' knowledge about and attitudes towards the use of models and modelling in learning science. *International Journal of Science Education, 24*(12), 1273–1292. https://doi.org/10.1080/09500690210163198.

Katz, J. S., & Martin, B. R. (1997). What is research collaboration? *Research Policy, 26*(1), 1–18.

Kinslow, A., Sadler, T., Friedrichsen, P., Zangori, L., Peel, A., & Graham, K. (2017). Connecting global climate change to a local ecosystem: A socio-scientific issue approach. *Science Teacher, 84*(7), 39–46.

McKenney, S., & Reeves, T. C. (2013). Systematic review of design-based research progress: Is a little knowledge a dangerous thing? *Educational Researcher, 42*(2), 97–100. https://doi.org/10.3102/0013189X12463781.

National Academies of Sciences, Engineering, and Medicine. (2015). *Science teachers' learning: Enhancing opportunities, creating supportive contexts*. Committee on Strengthening Science Education through a Teacher Learning Continuum. Board on Science Education and Teacher Advisory Council, Division of Behaviorial and Social Science and Education. Washington, DC: The National Academies Press.

NGSS Lead States. (2013). *Next generation science standards: For states, by states*. Washington, DC: The National Academies Press.

Peel, A., Sadler, T., Friedrichsen, P., Kinslow, A., & Foulk, J. (2018). Rigorous investigations of relevant issues: A professional development program for supporting teacher design of socio-scientific issue modules. *Innovations in Science Teacher Education, 3*(3). http://innovations.theaste.org/rigorous-investigations-of-relevant-issues-a-professional-development-program-for-supporting-teacher-design-of-socio-scientific-issue-units/.

Peel, A., Zangori, L., Friedrichsen, P., Hayes, E., & Sadler, T. D. (2019). Students' model-based explanations about natural selection and antibiotic resistance through socio-scientific issues based learning. *International Journal of Science Education, 41*(4), 510-532.

Romine, W. L., Sadler, T. D., & Kinslow, A. T. (2017). Assessment of scientific literacy: Development and validation of the quantitative assessment of socio-scientific reasoning (QuASSR). *Journal of Research in Science Teaching, 54*, 274–295.

Sadler, T. D. (2009). Situated learning in science education: Socio-scientific issues as contexts for practice. *Studies in Science Education, 45*, 1–42.

Sadler, T. D., & Zeidler, D. L. (2005a). Patterns of informal reasoning in the context of socioscientific decision-making. *Journal of Research in Science Teaching, 42*, 112–138.

Sadler, T. D., & Zeidler, D. L. (2005b). The significance of content knowledge for informal reasoning regarding socioscientific issues: Applying genetics knowledge to genetic engineering issues. *Science Education, 89*, 71–93.

Sadler, T. D., Foulk, J. A., & Friedrichsen, P. (2017). Evolution of a model for socio-scientific issues teaching and learning. *International Journal of Education in Mathematics, Science, and Technology, 5*(2). http://www.ijemst.net/issue/view/5000017771.

Schmidt, W. H., Wang, H. C., & McKnight, C. C. (2005). Curriculum coherence: An examination of US mathematics and science content standards from an international perspective. *Journal of Curriculum Studies, 37*(5), 525–559.

Schön, D. (1983). *The reflective practitioner: How professionals think in action*. New York: Basic Books.

Valli, L. (1997). Listening to other voices: A description of teacher reflection in the United States. *Peabody Journal of Education, 72*, 67–88.

Vo, T., Forbes, C. T., Zangori, L., & Schwarz, C. V. (2015). Fostering third-grade students' use of scientific models with the water cycle: Elementary teachers' conceptions and practices. *International Journal of Science Education, 37*(15), 2411–2432.

Williams, M. A., Friedrichsen, P., Sadler, T., & Brown, P. J. B. (2018). Modeling the emergence of antibiotic resistance in bacterial populations. *American Biology Teacher, 80*(3), 210–216.

Zangori, L., Peel, A., Kinslow, A., Friedrichsen, P., & Sadler, T. (2017). Student development of model-based reasoning about carbon cycling and climate change in a socio-scientific issues unit. *Journal of Research in Science Teaching, 54*(10), 1249–1273. https://doi.org/10.1002/tea.21367.

Zeidler, D. L. (2014). Socioscientific issues as a curriculum emphasis: Theory, research and practice. In S. K. Abell & N. G. Lederman (Eds.), *Handbook of research on science education* (pp. 697–726). New York: Routledge/Taylor and Francis.

Patricia J. Friedrichsen is a Professor in the Department of Learning, Teaching, and Curriculum at the University of Missouri. She teaches pedagogy courses in the undergraduate science teacher education program. Her research focuses on secondary science teacher learning across the professional continuum, from recruitment to professional development for classroom teachers. She draws upon a variety of theoretical perspectives including pedagogical content knowledge, teacher beliefs, communities of practice, and core teaching practices. In collaboration with the co-authors, she is currently researching core teaching practices and tools to support secondary teachers in using a model-based, socio-scientific issues approach to teaching biology. Pat's research work has been funded by the National Science Foundation. She has served as an elected board member for *NARST: A Worldwide Organization for Improving Science Teaching and Learning Through Research.* She was an Associate Editor for the *Journal of Research in Science Teaching* and serves on the editorial boards of the *Journal of Science Teacher Education* and the *International Journal of Science Education.* Prior to working in higher education, Pat taught middle and secondary biology for 13 years.

Troy D. Sadler is the Thomas James Distinguished Professor in Experiential Learning at the University of North Carolina at Chapel Hill School of Education. Sadler's research focuses on how students negotiate complex socio-scientific issues and how these issues may be used as contexts for science learning. He is interested in how issues-based learning experiences can support student learning of science and development of practices essential for full participation in modern democratic societies. He has also explored ways in which innovative technologies including virtual environments and gaming can support student learning. Sadler's work has been funded by the National Science Foundation, the Institute of Education Sciences, the US Department of Education, the Howard Hughes Medical Institute as well as local foundations and state agencies. He has served in elected and appointed leadership roles in the world's leading science education associations including *NARST: A Worldwide Organization for Improving Science Teaching and Learning Through Research* and the *National Science Teachers Association*. He is Co-Editor for the *Journal of Research in Science Teaching*.

Laura Zangori is an Assistant Professor of Science Education at the University of Missouri. She teaches and works with students ranging from elementary through undergraduate classrooms. Her research focuses on how to support teachers and their students in using systems models and modeling practices to understand biological systems and apply their scientific understanding to grapple with complex issues with environmental, societal, and political implications. Laura completed her PhD in Science Education at the University of Nebraska-Lincoln in 2015. Prior to her PhD work, she completed a M.S. in biology at the University of Illinois and spent 10 years in the pharmaceutical/medical device industry before shifting her career into biology teaching in 2007.

Chapter 7
Gamification of SSI's as a Science Pedagogy: Toward a Critical Rationality in Teaching Science

James P. Davis and Alberto Bellocchi

7.1 Introduction

The complexities of education in contemporary society may be viewed in terms of responses to vocational, citizenship, and individual goals (Winch 2004). Evidence of complexity may be interpreted from vocational contexts where high levels of knowledge, skills, teamwork, and analytical ability are required of contemporary workers. On citizenship, the maintenance of modern nation states requires citizens to be critical thinkers in a context of conflicting interest groups that adds to the uncertainty and complexity of social reality. At an individual level, people need to make choices about their own life goals, and the pathways for reaching those goals, within the context of their vocation and their being a part of society. This societal context of uncertainty and complexity provides grounds for an argument that educational goals need to include some form of *critical rationality* to be explicitly taught and learned in schools (Winch 2004). In science education critical rationality, commonly known as *critical thinking*, involves the ability to evaluate evidence and arguments from a range of scientific, social and ethical perspectives (cf. Simonneaux 2014; Winch 2004).

We suggest that the integration of gaming with authentic contexts involving socio-scientific issues ought to be an important part of science education because gaming enables spaces for student autonomy and critical rationality to become a possibility. The gamification of socio-scientific issues as a pedagogical approach in teaching school science is an emergent strategy. It may involve digital technologies (Psotka 2013), but typically includes a blended approach to gaming across virtual and real (physical) spaces. Daniel Dziob (2018, p. 2) points to the integration of

J. P. Davis (✉) · A. Bellocchi
Queensland University of Technology, Brisbane, Australia
e-mail: jp.davis@qut.edu.au; alberto.bellocchi@qut.edu.au

M. Evagorou et al. (eds.), *Science Teacher Education for Responsible Citizenship*, Contemporary Trends and Issues in Science Education 52,
https://doi.org/10.1007/978-3-030-40229-7_7

these spaces "as game mechanics, esthetics, and game thinking" may be applied to learning science within non-game or authentic contexts. In the context of this chapter we seek to illustrate this integration between elements of gaming, science and technology curriculum, and pedagogical strategies that embrace critical rationality as an essential objective.

Throughout this chapter, we make a case for the gamification of socio-scientific issues as a form of science pedagogy that may shift the focus of teachers and students toward a critical rationality in school science. We establish this case by explaining our own teaching experiences with gamification of SSIs in preservice teaching courses. Our focus in this chapter is around two questions:

1. What strategy and pedagogical tools may assist teachers to become producers of SSI games for implementation in school science contexts?
2. What are the possibilities for our strategy to contribute to critical rationality as an explicit aim in science pedagogy?

7.2 Motivation for This Work

As science educators the gamification of SSIs as a science pedagogy enables us to make science learning relevant and engaging for students. This outcome is achieved by not only connecting science to authentic contexts, but also enabling students to take meaningful gamified actions that are determined by the direction of student interest rather than being teacher directed. Gamification enables the teaching of *activist science* in both the exploration of science and in the meaning and application of science as a human endeavour. The project described in this chapter was initiated by the actions of Alberto in 2011 when he decided to introduce the notion of *Alternative Reality Gaming* (ARG) to a graduate preservice science teacher course focusing on curriculum planning. In the educational context of this study, alternative reality gaming involves an interactive storyline with the real world as its platform and uses a multi-media, multi-technology approach to deliver the story that can be altered by its players. It is played in real-time and can shift across virtual and real spaces where players solve story-based challenges and collaborate beyond their workgroups or classroom to take action in shaping reality. ARG is well suited for activist approaches to science education (cf. Bencze and Alsop 2014, also see Chap. 5 in this book).

Alberto It was around 2010 when I became aware of ARG through social media. I had been a gamer since my pre-teen years, mostly playing video arcade and console games. ARG caught my attention as something novel; the notion of blending a virtual world with real world experience was instantly seductive. Initially, following games mostly based in the US, I observed the basic game-play and then began reading (McGonigal 2011; Szulborski 2005) more about this novel and promising gaming genre, and subscribing to ARG forums like ARGNet (2018). Eventually I managed to enter a game close to the start of play. The game was set in the present and it involved a secret laboratory that was conducting human experiments under the guise of genetics health service. Characters within the game-world included staff working in the facility as scientists and assistants, and patients who had been subjected to suspicious treatments. The game characters interacted to unravel the storyline through the forum and players could participate by reading or contributing to discussion with the characters. Although virtual interactions were globally accessible from an electronic forum site where players and game characters interacted, real elements such as telephone calls and drop sites for material clues were only available to US players. For several weeks, bridging the time-zone divide, I eagerly logged in and contributed to the virtual dimension of play (players in the US received well-designed game artefacts in the mail, collected game clues at drop-zones, and also engaged in telephone calls). As the game was based in the US, none of the real-world dimensions were accessible by me. Nonetheless, play was consuming and the sense of anticipation each day to see how this unwritten story was unfolding become entrancing. The possibility of influencing the play by contributing through various online fora was intoxicating. I even created a fake Facebook profile for one of the characters (played by a player somewhere else in the world, whom I did not know) and started throwing in red herrings for others to deal with. There is no other game, in my experience, as unique as ARG. End-game took place early one morning (EST) in the online forum that supplied the main medium for solving game clues and facilitating character/player interactions. Anyone who missed that final forum was not privy to the game ending. Any player accessing the end-game script on the forum after the fact could not experience the sense of suspense and excitement generated by this in-the-moment aspect of game narrative.

After this experience and further reading, Jane McGonigal's educational ARG, *World Without Oil (WWO),* led me to consider more directly how I could use these concepts in my teaching. At the time, I taught two preservice science teacher courses for second and third year students: *Science, Technology, and Society,* and *Science Curriculum Studies.* The first included content related to SSI and social debates and implications for science (STS). In the second, preservice teachers learnt about high school science curriculum design. Forever the opportunist, I began exploring the possibilities of ARG for SSI/STS and in the development of engaging curriculum design in the second course. Aside from close parallels between ARGs like WWO and SSI/STS curricula, these games involved (at the time) extensive use of Web 2.0 technologies and social media; technologies I was seeking to embed in my teaching. Despite some enthusiastic claims about current generations, experience taught me that few of my preservice teachers really knew about these technologies, let alone how to adopt them meaningfully in teaching.

These explorations led to writing an article focusing on the possible integration of ARG into school and preservice science teacher education (Bellocchi 2012).

Some of the structural features of ARG surfaced complex ethical questions for my university teaching, which informed suggestions for school teaching. In some ways, I felt that the idea (only a short 7 years ago) was too far ahead of its time: Better access to social media technology in schools and university would be required; policies about use of software would need revision in the university; preservice teachers were very unfamiliar with ARG, location-based gaming, and mixed-reality.

In James, I found a willing participant. Not only did he entertain my game-based exploration of curriculum in the science curriculum course, but he adopted the ARG game framework in the design of his science curriculum assessment task. This became one of the examples I reported in my article (Bellocchi 2012). Having completed his doctoral studies under my supervision, chance would have it that James took over the teaching of my courses as I embarked on a 3-year research fellowship. His enthusiastic uptake of the gaming ideas I had offered his class were extended when he became the STS and curriculum studies lecturer. The year was 2016 and, given the fast pace of gaming progress, mixed-reality, ARG, and location-based games had received greater public attention.

James As a preservice teacher in Alberto's course I found the notion of ARG as offering a provocative approach to planning that connected well with context. It also provided an opportunity for me to make choices in the direction of my learning. With that opportunity I took a risk in refining a basic framework suggested by Alberto for developing a gamified scenario to situate a plan for teaching a secondary biology class. This was the start of my interest in the gamification of science education.

7.3 Theoretical Framework

Alberto In 2012, under the leadership of Steve Ritchie, a project focused on emotion and SSI in middle-school science was successfully funded. Based in Sadler's (2009) definition, I understood SSI to be social issues that relate to science and contain an ethical or moral dimension (e.g., food shortages, global warming, oil crisis) that could be explored from a scientific viewpoint or having embedded scientific/technological dimensions, problems, or prospective solutions. This view of SSI was also based in some foundational concepts from the STS literature. My approach to ARG, particularly in the STS course I taught, was framed against these scholarly traditions. Some ARGs (e.g., WWO) I was studying at the time were very closely aligned with the SSI/STS academic literature. For instance, WWO explored what social and environmental implications would arise once the last drop of oil had been consumed. Games like this offer ideal models for exploring ethical and moral dilemmas of, for example, government decisions on policies around the oil crises, or climate change a two examples. My preservice teachers developed assignments based on these principals and their own further research.

James By mid-2016 I was teaching and coordinating some of Alberto's courses, and by that time my approach to SSI's was strongly informed by Simonneaux's (2014, p. 104) "Socially Acute Questions" (SAQs). The notion of SAQs as a means for defining SSIs requires the science educator to embrace pluralism in epistemic approaches by engaging in risk and controversy, drawing upon contextualized data and accepting the construction of knowledge as multi-layered and distributive (Simonneaux 2014). For me, this requires an appreciation of non-scientific understandings of SAQs including positions informed by ethics, values and ideologies as well as conventional scientific thinking.

The adoption of SAQs situates my teaching toward inter-disciplinarity between the sciences and the humanities where preservice teachers are encouraged to adopt critical approaches to decision making and activist approaches in the production of outcomes in their course assessment. Simonneaux (2014) described this approach to SSI's as *hot* educational objectives where philosophical values, scientific and political citizenship could be enacted as the vehicle for learning science as an enacted practice. This theoretical approach to understanding SAQs and activist interdisciplinary thinking was a feature of my explicit teaching with preservice science teachers in the first half of their semester-long course.

7.4 Design Case

The design of this study is grounded in a cycle of design thinking and reflective practice where Alberto and I were participant/researchers exploring our own understanding and learning about the gamification of SSIs, and how this may be best taught in an initial teacher education (ITE) environment. This design has reflected our approach to both teaching and learning as teachers through ongoing modifications to our practice and professional dialogue over time. The challenge at the heart of this design process is to develop a pedagogical strategy and practical tools to enable teachers to conceptualise a SSI game and to develop the game for implementation in school science contexts.

7.5 Methods of Data Production

Our methods of data production in this study comprise teaching reflections and artefacts collected through our work in designing and delivering teaching. For Alberto, his journal article (Bellocchi 2012) and a conference proceedings paper (Lloyd et al. 2012) document research activity and data analyses related to the ARG model used in his teaching. Data sources include his teaching reflections, Wiki, blog, and discussion board extracts, interviews with preservice science teachers, and assessment items from two courses. For James, his teaching reflections include reflective notes, the assessment item that he wrote as a student of Alberto, and his PowerPoint materials and planning notes from the courses he taught.

7.6 Findings

Our findings from this design thinking process are summarised here as three episodes of development: Alberto's teaching in 2011, and James's teaching in 2016 and 2017. The first two episodes show the unfolding of an ARG architecture to design the process of a game and the final episode combines the ARG architecture with a unit planning framework to integrate the science curriculum with technology, society, learning experiences and assessment methods.

7.6.1 Episode 1: Immersion into the World of Alternate Reality Gaming

Alberto Our gamification approach (i.e., the one James was exposed to as my student) drew directly on the ARG framework (Bellocchi 2012). *TINAG, This is Not a Game*, is the foundational principle of ARG developed originally by Szulborski (2005). For a game to proceed, players willingly suspend their knowledge of being in a game. ARGs also require a *puppet master* (i.e., game staff); a role that I took on when sharing the idea with my classes. Puppet masters are responsible for creating the *curtain;* a metaphor used to capture the separation of players from the puppet master's true identity as the architect of the game. Curtains could include websites designed to look like some artefact in the game. For example, in a game involving some medical theme, the curtain could be a website designed to look like that of a hospital at the centre of the game. The puppet master's role is to create an environment that upholds the TINAG principle by means of a curtain. This includes playing a major character or a lead character who initiates play, which begins with a *rabbit-hole.* Drawn from Lewis Carroll, rabbit-holes are the entry point of the game that allow players to become lost in a labyrinth of possibilities. Rabbit-holes could be online fora where clues that lead to a website (e.g., the curtain) are planted, or a QR code on a poster that leads you to a game clue. A moment's perusal of ARGnet will reveal the multiple and unique ways in which a rabbit-hole is created. Given the potentially disorienting nature of ARG (rabbit-holes can meander somewhat), *trailheads* are used to keep players on-track with the intended and emerging storyline, game clues, and player discoveries. A trailhead is also a deliberate clue left by the puppetmaster to keep the game play advancing. Online forums offer great locations (i.e., virtual spaces) for trailheads.

I used a WordPress™ page as my rabbit-hole and the university online learning management system (LMS), Blackboard™, discussion board as the trailhead.

A video with me dressed as an older farmer introduced the SSI problem, and during face-to-face teaching, my preservice teachers then explored the scenario. Challenges were posted on the discussion board that forced my preservice teachers to engage with a range of Web 2.0 tools in the LMS. One involved locating an engaging resource in the library's curriculum section, photographing it, and then creating a Wiki page for sharing with other members of the class. In addition to posing the challenge of learning about Wiki's (i.e., functionality, purpose etc.), this task also demonstrated how preservice teachers could find engaging ways for sharing resources; modelling practice they could adopt in future classes. Each challenge could only be completed through self/peer-teaching of the technology involved (e.g., blog, wiki). This strategy again involved my modeling of teaching practices that could be adopted with high school science classes.

7.6.2 Episode 2: Developing an ARG Architecture

James In 2016 I was presented with an opportunity to teach an undergraduate course for preservice science teachers using a STS approach for teaching around SSIs. The assessment for that course involved the development of a course plan in small groups, followed by the presentation of a short lesson from the course plan on an individual basis. I identified this as an opportunity to introduce the ARG strategy that I had previously applied in my work with Alberto in 2011. To illustrate the ARG strategy I developed a flowchart that I have called an ARG Architecture as shown in Fig. 7.1.

The initial ARG architecture shown in Fig. 7.1 shows the basic features, which may play out through an iterative process depending on the duration and complexity of the game. This architecture consists of the following features:

(a) *Puppet Master.* The role of the puppet master is to plan, set rules and manage progression of the game, knowing that players will take the game in different directions. The teacher as puppet master is a good way to start because the future options for game direction can be managed so that various learning objectives may be embedded within the pathways that students choose to explore.
(b) *Rabbit-hole.* The start point for the game is the rabbit-hole, which should be playful, fun, and a little eccentric, as a way of enticing students to get involved and stimulate their imagination. The puppet master could introduce the rabbit-hole with a role play, which may involve brief purpose made video. In the game I developed called *Cool and Covered* my goal was to promote design activities to develop skin cancer prevention solutions. I used a cartoon character called *Sid the Seagull* who is well-known in Australia as the mascot for a skin cancer prevention campaign. I was able to use online videos made by Cancer Council Australia (2018) with this ready-made character to introduce the SSI of skin cancer prevention. I then extended beyond the introduction by Sid the Seagull

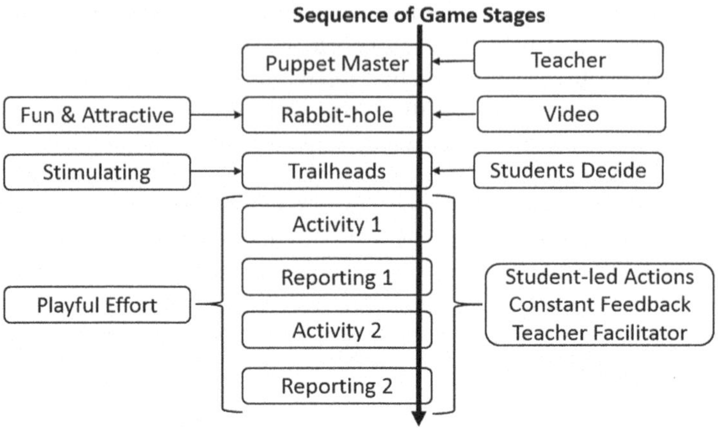

Fig. 7.1 Initial ARG Architecture

with a collection of images to introduce issues of design and shading in public spaces, the design of clothing and cultural practices in Australia that challenged the skin cancer prevention message. These challenges provided opportunities for game participants to appreciate that everyday decisions are not made on scientific grounds, and that the interconnections between what people might know and how they may behave need to be understood from inter-disciplinary perspectives.

(c) *Trailheads.* With my preservice teachers I provided a very open-ended approach with the trailheads, which were revealed sequentially as (1) explore the problem, (2) design a solution, and (3) tell your community. For example, in the trailhead of *explore the problem* each team was to explore one of the sub-issues they identified in the rabbit-hole and come up with a justification and a plan for taking action. I purposely allowed them to explore and find a sub-topic of their choice because this situated to gain an appreciation of the broader topic while selecting one as a specific area of interest. At that point they would report back to the puppet master, and receive points for that stage of the game followed by instructions for the next stage, which was *design a solution*. From an educational perspective these points of interaction with the puppet master are stages of formative assessment, feedback, reward and further guidance. The degree of detail in guidance for each stage will vary with the students and the context, but opportunities for students to maximise creative input and autonomy should be a priority. The number and sequence of trailheads could also be varied and should enable an iterative cycle of engagement throughout the game.

More structure in the storyline would be necessary with children in primary or secondary school contexts. Greater emphasis would be required on the characters such as Sid the Seagull with opportunities for students to develop their own characters within the game. For example, specific problems could be identified such as helping Sid to design a more fashionable hat, or helping Sid improve the shade areas in the playground. To shape the planning of ARGs to this level of analysis the earlier architecture in Fig. 7.1, was further developed in to a planning framework, as discussed in the following section.

7.6.3 Episode 3: Planning Framework for an ARG

James's Teaching 2017 In 2017 I taught another course to undergraduate preservice teachers entitled *Science Technology and Society* (STS), and developed a more detailed approach towards integrating the game with the formal school curriculum. The first half of the semester focused on preservice teachers understanding the principles of STS education and the nature of socio-scientific issues. As part of the reading and teaching, students engaged with the idea of SAQs and interdisciplinary approaches to teaching science (cf. Simonneaux 2014).

Students developed and explored a SAQ that involved getting out of the classroom and interviewing a scientist about the issue. In this context I adopted a broad notion of the term *scientist*, which enabled preservice teachers to explore issues by engaging with people such as ecologists, paramedics, nurses, veterinarians, science related policy makers, geologists, medical practitioners, research scientists, environmental activists and so on. Some of these people were part of the preservice teachers' existing networks, while other scientists were engaged by preservice teachers expanding their networks. This expansion of networks was empowering for preservice teachers as it gave them self-confidence in talking to people outside of their field and it also broadened their own conceptualization of who counts as a scientist. As an example, I recall one primary school preservice teacher who managed to get an invitation to a Department of Health seminar addressing the issue of medical marihuana. That preservice teacher was able to discuss the issue with medical practitioners and policy makers. I was very impressed with the activism of my preservice teachers, evident in the way they acted on the opportunities provided to them.

The second part of this unit involved preservice teachers developing a plan to teach the issue at the focus of their SAQ and in this part of the course I introduced the gamification framework for course planning. I taught this framework over 4 weeks where I explained the reasoning of my planning, demonstrated it with a real issue, and then supported preservice teachers to develop their own plan for a gamified course design in a school context. The conceptual overview of the ARG and course planning framework is shown in Fig. 7.2.

The structure of Fig. 7.2 shows the planning process I undertook in developing a sample game for my students. The sequence of the game flows across the top row from left to right. Consistent with the earlier structure, the game is initiated with a rabbit-hole that aims to spark student interest. This starting point leads to trailheads giving students an opportunity to choose a topic or a direction for their initial activ-

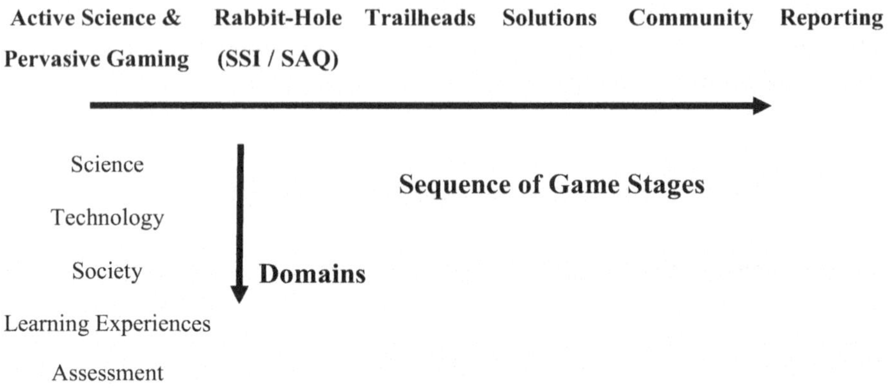

Fig. 7.2 ARG and Course Planning Framework

ity. From here students can explore, inquire, develop and engage with others in the classroom, beyond the classroom or in their after-school community.

I explicitly connected the TINAG rule, introduced earlier by Alberto, with the notion of activist science because the SAQs are real issues about which students could take real action. To enhance this activist science/TINAG factor I added community engagement as one of the cyclical stages of the game planning framework. Taking the game out of the classroom was therefore an important element that students could develop and report about. As students report back to the puppet master it is possible for new trailheads to be explored making gamified learning an iterative experience that jumps back and forth across the stages described in the first row of Fig. 7.2.

To scaffold the planning process for my preservice teachers I developed planning questions for each stage of the game across the domains of science, technology, society, learning experiences and assessment. A summary of these questions is shown in Table 7.1. As an example, to design the *rabbit-hole* stage for the *science* domain I asked questions such as *what is the SSI/SAQ at stake here?,* and *What key science concepts are the focus of the game?* These questions were designed so that preservice teachers could work methodically through the ARG and Course Planning Framework. This process shaped their thinking toward a cohesive summary of the salient features of their game in the context of teaching, and the specific curriculum objectives. This level of thinking and planning, in response to previous feedback, was necessary to ensure preservice teachers would be prepared to respond to future students, and to guide the preservice teacher as both teacher and puppet master. It also enables planning of logistical limitations that may restrain student choices such as the type and amount of materials and equipment available in any particular school.

Cool and Covered: Designing Out Skin Cancer *Cool and covered: Designing out skin cancer* was the topic I used to illustrate the gamification of teaching in science

Table 7.1 Planning Questions

Active Science & Pervasive Gaming	Rabbit-Hole (SSI/SAQ)	Trailheads	Solutions	Community	Reporting
Science	What is the SSI/SAQ at stake here? What key science concepts are the focus of this game?	What science inquiry topics are you directing students toward?	What solutions are feasible for students to develop?	What opportunities are there to build community involvement?	What avenues are available for students to report their outcomes?
Technology	What technologies are involved in the SSI?	What science technologies are involved? What re-design opportunities are possible?	What materials and technologies will students require to implement solutions?	How will students engage the community with their technologies?	What changes to technologies are possible as outcomes of the game?
Society	What are the social, ethical, emotional, economic aspects of this issue?	How might these social, ethical, emotional, economic aspects be addressed?	What social solutions are feasible for students to implement? How could community, school, family contact be implemented and evaluated by students?		How could social impacts be evaluated and then reported by students?
Learning experiences	Develop a rabbit-hole to entice student participation and introduce the topic and context of the game. Determine pedagogical technologies you will use.	What are the possible pathways students could take in terms of experiences in science inquiry or technology design? What are the resource and time constraints? How will student choices be scaffolded?		What are the possible community links that students may want to access and how can these be incorporated into science inquiry, technology design solutions and/or science education beyond the classroom?	In what format will students report? What scaffolding is needed to enable this reporting? What ICT's do students have access to for the purposes of disseminating reports?
Assessment How will each part of the game be scored & linked to assessment?	Define and explain the goals of the game? How does a student win the game?	What guidance, feedback and scaffolding might you need to provide at this point?	What guidance, feedback and scaffolding might you need to provide at each of these stage? For each stage, should the student cycle through another trailhead?		

for my course in 2017. The game I modelled was aimed at the Year 8 and 9 Australian Curriculum (for students aged 13–14 years) where science topics such as energy, materials and cell biology could be integrated with the SAQ and related social and ethical issues. The SAQ at the focus of this game was:

> *How could we design technologies in our everyday life to eliminate the incidence of skin cancer?*

In Australia, skin cancer continues to be a leading cause of death, despite many years of public education campaigns. This SAQ is a question being investigated within the QUT School of Public Health in collaboration with the Creative Industries Faculty. As the rabbit-hole for this game, I introduced the topic with a video that I embedded with supporting assessment documents in a Padlet™ page. Padlet™ is a university supported cloud-based technology designed to elicit online interaction between students. The rabbit-hole is quite a brief presentation and led preservice teachers to a number of trailheads as different pathway choices that could be explored. I located links to the trailheads in my Padlet™ and these links took students to different pathways that they could develop further in Microsoft Sway™ (a digital story telling app). To play the game, preservice teachers created a copy of their selected trailhead in Microsoft Sway™ and this became the online vehicle for recording gaming events as the story of their game-play unfolded over time. When students engaged with offline activities they were able to record these events through images, notes, reports and so on within their Sway™ document.

Within these trailheads preservice teachers were guided toward sun protection *technology design choices* by encouraging them to think about potential technologies such as sunscreen, hats and clothing materials, built shade, and types of building materials that could provide cover by blocking or reflecting UV light. Importantly the design element of these technologies would not only account for science and technology considerations but also perspectives reflecting social, ethical, aesthetic and emotional dimensions to successful design. I provided this level of guidance to enhance their appreciation of the complexity of this problem. For example, hats have been around for centuries, so it is not a lack of hats that prevents people from wearing them. Design choices needed to reflect broader cultural issues so that student designs of protective technologies would promote end user engagement. At practical level, I also needed to ensure that choices were aligned with the resources I have available in the Faculty's science lab, which reflects the resources of a typical secondary school lab in Queensland, Australia.

In the context of this game topic I am able to demonstrate science inquiry possibilities such as testing the impact of ultraviolet (UV) light on different analogues for skin cells. For primary school contexts I demonstrate the use of beads that change colour when exposed to UV light. For secondary students, we are able to incubate non-pathogenic varieties of normal flora from human skin such as *Staphylococcus epidermidis*. For this context I had preservice teachers make and test sunscreens with recipes obtained from the internet compared with commercial sunscreens. I have also had preservice teachers test the protective quality of different clothing materials, based on material type and colour. Other suggested examples could include shade surveys on campus, or addressing issues such as the poor design of our local bus shelter at the front of the university. These examples of science and technology inquiry may also involve technology design solutions and social research about aesthetics or localised cultural practices that could be developed as further trailheads.

In this particular course my focus was on the planning of the game for a school context, and for this reason my preservice teachers did not fully implement these suggested activities. However, in other courses, I have implemented these exact activities, and on this basis I have tested my capacity as a teacher to engage preservice teachers in multiple inquiry and design activities at the same time. There is definitely a level of self-confidence that a teacher needs to develop to be able to deliver such bespoke science learning experiences in the one classroom. This experience reinforces my commitment to ensuring preservice teachers are equipped to plan in detail, by effectively *gaming-the-game* before presenting it in class as the puppet master.

7.7 Discussion

Throughout this chapter we have presented our experiences with our design and reflective evaluations of alternative reality gaming (ARG) as an approach to planning and teaching science through the lens of socio-scientific issues (SSIs) or socially acute questions (SAQs). This study is an historically situated reflective account of our own teaching experiences. Our presentation of these accounts evidences a chronological and cumulative achievement of an ARG Planning Framework that has unfolded through our teaching experiences. The design has unfolded contemporaneously with our teaching from the historical sequence, and periods of discontinuity, of our shared teaching experiences and teaching interests over the past 7 years. The strength of our ARG planning model is in the way it is shaped by our professional growth as science educators and science education researchers. Alberto's original passion for gaming, for integrating technology with his teaching, and for making science education engaging provided the impetus for the initial growth of this teaching focus. In contrast, James was not a gamer, but could see the value in the ARG concept for engaging students in the complexity and uncertainty of socio-scientific issues. The current location of this project enables us to propose an ARG strategy, a small collection of planning tools and experientially grounded advice. In response to our first research question we offer this chapter to assist teachers who may choose to become ARG producers, or choose to engage their students in producing ARGs.

There is much more to be done in the gamification of SSIs as a science pedagogy, particularly in the development and integration of digital technologies such as augmented and virtual reality. This will only become possible as more teachers become involved in game production, alongside software designers so that the expertise of educators and coders can be fully integrated. Such interdisciplinary approaches also need to include the education of teachers so that science teaching is led with SSIs or SAQs. To achieve the hot end of the activist science spectrum, teachers need to be comfortable in teaching around and through controversy, as well as the uncertainty and complexity that arises when adopting multi-disciplinary perspectives with students. In addition, and consistent with constructivist approaches to learning, teachers could invite students to create their own games with or without the support of game designers. Considerations that students would need to make as puppet masters would support their critical thinking, which may be enhanced if the games designed by students are played by peers. As peers engage in game-play and contribute ideas and solutions to problems about, for example, an environmental disaster game, the puppet master students would have to make choices in how to steer the game through their interactions with players. Consideration of the environmental problem, including science and social issues would be required in their critical decision making. Teachers could ask classes to reflect critically on the SSI after a game has been played. Even if students were to play existing games, such as WWO, this kind of reflection could support a critical pedagogy.

What we are suggesting here is the need for teachers to be educated and confident in the practice of critical rationality (Winch 2004). Simonneaux (2014) suggests that the ways in which SAQs are addressed in science teaching typically reflects the rationality of the teacher and that this may also vary from topic to topic depending on the teacher's interests. This is an interesting observation by Simonneaux (2014) because Winch (2004) suggests that critical rationality is context specific and may also be variable in terms of how strong a teacher may want to exercise their own autonomy for critical thinking. For these reasons, while gamification opens the possibilities for students to exercise their autonomy, it is not the only factor that is influencing the degree of critical rationality possible in the classroom. As the puppet master the teacher's sense of autonomy and appetite for critical rationality may constrain the full potential for games to produce a science pedagogy with strong elements of critical rationality. We point to the earlier suggestion of handing over the role of puppet-master to students. By having students' create their own games, they would be engaging with the science while removing the teacher's sense of autonomy as a potential constraint. In this sense the game truly becomes the science pedagogy.

7.8 Implications for Teaching and Research

Our approach in the gamification of SSIs as a science pedagogy has de-emphasized the use of ready-made digitalized games, and instead focused on the application of game thinking to course planning and the integration of common technologies, curriculum, game elements, and authentic socio-scientific contexts. The limitation of this approach is that digital-based gamers may be disappointed with what we offer, but we do welcome greater collaboration between game coders and educators so that digitized games are meaningful and impactful from curriculum perspectives.

The strength of our study is its emphasis on planning across the stages of the game that are integrated across science curriculum, technology, diverse societal perspectives, and the pedagogical features of learning experiences and assessment. The proposed ARG Planning Framework articulates a start point for further research and teaching practice that may stimulate teacher interest in the development of new gaming opportunities in the teaching of SSIs.

A further dimension to this study has been the integration of critical rationality, which is highly dependent on the teacher's role as the puppet master, but could be enhanced by engaging students in that role. We have suggested that gamification of SSIs/SAQs does increase the possibility for student autonomy and in this sense gamification provides a distinct structural possibility for critical rationality to be actively pursued as a pedagogical aim. In light of the current global challenges to science around notions of what constitutes valid knowledge, the time for high quality education in the practices of critical rationality have never been so important to a cohesive and forward thinking global society. Clearly a science pedagogy of gamification as a vehicle for critical rationality in science education, is an area that is ready for further technological development, teacher co-production, and practice-focused research.

References

Alternate Reality Gaming Network [ARGNET]. (2018). Retrieved from https://www.argn.com on 09 July 2018.

Bellocchi, A. (2012). Practical considerations for integrating alternate reality gaming into science education. *Teaching Science, 58*, 43–49.

Bencze, L., & Alsop, S. (Eds.). (2014). *Activist science and technology education*. Dordrecht: Springer.

Cancer Council Australia. (2018). http://www.sunsmart.com.au

Dziob, D. (2018). *Board game in physics classes – A proposal for a new method of student assessment*. Research in Science Education. Advance Online Publication. https://doi.org/10.1007/s11165-018-9714-y

Lloyd, M., Mukherjee, M., & Bellocchi, A. (2012). Re-imagining STEM: Peer scaffolding ICT in initial teacher education. In *2nd international STEM in education conference*, Beijing, China, pp. 131–138.

McGonigal, J. (2011). *Reality is broken: Why games make us better and how they can change the world*. GB: Jonathon Cape London.

Psotka, J. (2013). Educational games and virtual reality as disruptive technologies. *Educational Technology & Society, 16*, 69–80.

Sadler, T. D. (2009). Situated learning in science education: Socio-scientific issues as contexts for practice. *Studies in Science Education, 45*, 1–42.

Simonneaux, L. (2014). From promoting the techno-sciences to activism – A variety of objectives involved in the teaching of SSIs. In L. Bencze & S. Alsop (Eds.), *Activist science and technology education* (pp. 99–111). Dordrecht: Springer.

Szulborski, D. (2005). *This is not a game: A guide to alternate reality gaming*. Morrisville: New Fiction Publishing.

Winch, C. (2004). Developing critical rationality as a pedagogical aim. *Journal of Philosophy of Education, 38*, 467–483.

James P. Davis is a lecturer in STEM education, teacher leadership and entrepreneurial thinking in the Faculty of Education, Queensland University of Technology. His research focuses on the interplay between emotion and social cognition in learning using approaches such as ethnomethodology, phenomenology and micro-sociology. James views gamification in teaching science as an important, but underutilized, approach to promote the emotional and social engagement of students with learning science.

Alberto Bellocchi is an Associate Professor and Principal Research Fellow in science education, within the Faculty of Education, Queensland University of Technology. His program of research over the last decade has focused on school students and pre-service teachers' emotional engagement with science. Drawing on micro-sociology and theories of emotion, social bonds, and interaction, he studies the complexities of lived experiences involved with learning science.

Chapter 8
Science Teachers as Proponents of Socio-Scientific Inquiry-Based Learning: From Professional Development to Classroom Enactment

Rachel Cohen, Eran Zafrani, and Anat Yarden

8.1 Introduction

In many aspects, twenty-first century science constitutes a dominant and pervasive factor in people's lives, the extent and reach of which extend well beyond the professional scientific community to influence societies, the environment and the lives of individuals. The pervasiveness of scientific and technological advances is exemplified by continuing discussions in the public sphere that focus on current issues such as climate change, access to clean water, food shortages, genetic modification, and other critical issues that mandate all citizens' critical attention. Since these topics raise questions that relate both to their scientific and social dimensions, they have been termed as socio-scientific issues (SSI). Because of their immediate effect on society, SSI necessitate increased public awareness and practical involvedness from all citizens. Without citizens' active participation, the safety of our lives, societies and the environment may be jeopardized (Bencze and Carter 2011). Therefore, knowledge of and about the connections between science and society is a necessity for all citizens – scientists and non-scientists alike. Moreover, when dealing with SSI, students are required to engage actively and responsibly with science and to offer scientifically informed solutions where social implications appear to exist with the purpose of working toward providing a safer world (Aikenhead 2005; Kolstø 2001; Zeidler et al. 2005). Advancing the notion of scientific understanding for active and responsible citizenship is, therefore, a central concern of the science education community.

As is the case with other instructional demands, teachers have an integral role in answering the mandate to support the development of students' ability to engage actively and responsibly with SSI. Teachers are therefore key agents in any attempts

R. Cohen · E. Zafrani · A. Yarden (✉)
Department of Science Teaching, Weizmann Institute of Science, Rehovot, Israel
e-mail: rycohen@bezeqint.net; eran.zafrani@weizmann.ac.il; anat.yarden@weizmann.ac.il

M. Evagorou et al. (eds.), *Science Teacher Education for Responsible Citizenship*, Contemporary Trends and Issues in Science Education 52,
https://doi.org/10.1007/978-3-030-40229-7_8

to implement SSI instruction in classrooms. However, since school science has traditionally separated science content from the social implications of science, teachers also require adequate preparation for these attempts to succeed (Bencze and Sperling 2012; Lee et al. 2012).

This chapter begins with a review concerning the literature on SSI and how the implementation of SSI into classroom practice could facilitate increased civic participation. This discussion is followed by a presentation and rationale for an educational approach that integrates SSI with inquiry-based learning as a practical and pragmatic approach to promote active citizenship in science education. This approach was developed in the course of the EU-funded PARRISE (Promoting Attainment of Responsible Research and Innovation in Science Education) project and was termed socio-scientific inquiry based learning (SSIBL) (Levinson and PARRISE-Consortium 2014). Subsequently, the role of teachers' professional development for preparing to teach science in the context of SSIBL is discussed and a model for a teacher professional development course designed to familiarise teachers with SSIBL and to prepare them for implementing this approach in their classrooms is explicated. To make sense of this model, the experiences of two case study teachers who participated in a course that was designed based on this model are described.

8.2 Theoretical Framework

8.2.1 Socio-Scientific Issues and the Role of Students as Active Citizens

While recent scientific and technological developments have positively contributed to our overall wellbeing (consider for example the longer human life span), they also represent social complications and new risks that individuals and communities need to learn how to deal with. These include, for example, impacts of genetic interventions both in health management and in agriculture that have yet to be systematically analysed; certain human populations that are at risk of having low access to clean water and diminished food security; and the increased production of greenhouse gasses and the subsequent increase in earth temperature that has raised anxieties about the future wellbeing of individuals, society, and the environment. All of these issues share a unique characteristic in that they involve one or more cultural, ethical, moral, economic or political concerns and therefore pose an inherent social significance. Because of their social complexity, the inclusion of SSI teaching in science education has a potential for promoting competencies essential for active citizenship (Berkowitz and Simmons 2003). It is not surprising then that science education researchers, as well as various national curricula around the world, call for increased attention to SSI in science education (Bencze et al. 2012; Hodson 2003; Levinson 2010).

In research, efforts for addressing SSI in science education have been formulated in terms of developing students' scientific literacy (Aikenhead 2005; Dos Santos 2009; Hodson 2003; Kolstø 2001; Sadler and Zeidler 2009; Zeidler et al. 2005). The definition of scientific literacy is well known to be subjected to multiple and varying interpretations and not all of them reflect on SSI. In fact, when looking at the types of scientific knowledge that are emphasised inside schools, the educational pendulum is still leaning towards what Roberts (2013) described as Vision I scientific literacy, meaning a scientific literacy that mainly focuses on canonical laws and theories; while Vision II, which focuses on the role of science and scientific knowledge for everyday life, is still scarcely exercised (Bencze and Carter 2011).

Though still not representing the status quo in school science practice, the type of scientific literacy contained within Roberts' Vision II has gained considerable recognition over the years and current formulations of scientific literacy emphasise the social dimension of science as pertaining to proficient scientific literacy (Hofstein et al. 2011). Within this school of thought, several contemporary and prominent formulations explicitly emphasise the role of science education as a tool for empowering students to react to SSI as responsible citizens. For example, in their proposed framework for SSI instruction, Zeidler et al. (2005) argue that proficient scientific literacy should provide students with opportunities to make informed decisions regarding SSI, thus making them active actors who can negotiate and resolve criteria about SSI. In Israel, this performative notion of science literacy for informed decision making also had a curricular response. In 1992, the Israeli Ministry of Education appointed a committee to observe the country-wide status of science education (Israeli Ministry of Education 1992). As a response to earlier reflections on the interactions between science and society that are rooted in the STS movement, the outcome of this committee's work, the 'Tomorrow 98' report, presents a more engaging pedagogy for science teaching through placing a special emphasis on social aspects of science and science laden technology. The report makes an argument for science as an exercise connected to everyday life in a way that mandates students' decision making regarding current science and technology issues (ibid.).

A more explicit focus on students' active participation as citizens affected by science can be found in Hodson's (2003) seminal paper in which he emphasised the importance of students' action as integral to the promotion of scientific literacy. He argued that we need to consider scientific literacy as a concept that should promote students' 'capacity and commitment to take appropriate, responsible and effective action on matters of social, economic, environmental and moral-ethical concern' (Hodson 2003, p. 658). A similar argument is presented by Dos Santos (2009) who proposed a humanistic perspective on science literacy that emphasises students' social action for the common good. He argued that science education should reflect on issues of social injustice and inequity, and consequently be aimed at the transformation and creation of a better society. Another clear view of scientific literacy for civic change can be found in Aikenhead's (2005) position which highlights social responsibility and students' practical actions. In all these publications, the conceptualisation of scientific literacy shares an objective and a vision of science education that encourage students to make appropriate decisions and to take participatory action on issues that involve science and society.

Several studies have shown the utility of engagement with SSI in terms of advancing the role of students as active and responsible citizens. For example, Roth and Lee (2004) investigated an educational programme which involved students learning science through participation in an environmental project set in their community. When students acquire knowledge by contributing to their community, they argued, it can pave the way to lifelong participation and learning of science. Barton and Tan (2010) argued that students' participation in a science project that includes a component of activism for the benefit of their community afforded participating youths to frame themselves as 'community science experts' thus providing them a sense of empowerment to act as concerned citizens. Similar results were presented by Zafrani and Yarden (2017), who showed how an activity structured around an SSI that deals with global hunger promoted student motivation and willingness to act in order to resolve this issue by means of scientific inquiry and humanitarian work. These and other studies converge on the conclusion that when science education is embedded in community contexts, it can be meaningful in terms of students' participation both in school as science learners, as well as in their communities as active citizens.

8.2.2 Socioscientific Inquiry Based Learning

Taking a more active stance on scientific issues requires students to make the transition from discussing SSI in theory to making informed decisions and proposing concrete solutions that address the examined issue and result in some kind of change. Because these issues originate from a dilemma informed by science, students' solutions ought to rely on formulation and interpretation of scientific evidence (Bencze et al. 2012). As well, for students to be able to propose informed solutions they are required to understand how scientific knowledge pertaining to the dilemma was constructed (Hodson 2003; Walker and Zeidler 2007). Furthermore, the combination of scientific and social dimensions, which together formulate ongoing controversies, raise many open questions and provide valuable possibilities for scientific inquiry that are embedded in real-world issues (Sadler et al. 2007).

Inquiry-based learning, in various adaptations, was previously discussed in connections with SSI instruction. For example, Walker and Zeidler (2007) designed a learning unit that challenges students to engage in web-based inquiry about SSI in the context of genetically modified organisms in agriculture and to apply their understandings in a discussion about policymaking regarding this issue. Students who engaged with this unit therefore applied knowledge attained from inquiry towards civic decision making. Sadler et al. (2007) also utilised a web-based learning environment for inquiry into the issue of water pollution. Bencze et al. (2012), documented the works of teachers who directed both open end correlational inquiries and web-based inquiries in the context of different SSI.

One way to consider inquiry-based learning as a means for contextualizing SSI instruction is through socio-scientific inquiry based learning, an approach that integrates the teaching of science using socio-scientific issues (SSI) with inquiry-based learning. This combination of contextual engagement with SSI and application of knowledge through scientific inquiry processes is therefore argued to increase students' understanding of SSI in a way that will allow them to enact their civic responsibilities and to propose solutions that are accountable to scientific theory and knowledge (Levinson and PARRISE-Consortium 2017).

8.2.3 Teacher Preparation and Learning to Teach SSIBL

Despite curricular mandates and extensive representation and academic justifications in research, implementation of education regarding SSI in schools has been limited (Levinson and PARRISE-Consortium 2017). When the SSI approach is adopted as a legitimate object of reflection in the classroom, the instruction is mostly constrained to a presentation of the social dilemma, with no attempt to promote students' meaningful participatory engagement or action (Bencze et al. 2012) which makes the implementation of SSIBL more difficult. Since teachers are the primary intermediaries for the curriculum, and since they are in proximity to the instructional situations – they choose how to implement curricula and how to work with students – they are considered key actors in any attempt to promote engagement with SSI in the classroom, both as an instructional practice or when contextualized as inquiry (Bencze and Sperling 2012).

By allowing SSI and IBL to meet, SSIBL introduces new concepts to science teaching that are novel to many teachers. In order to be able to conduct lessons that focus on SSIBL, teachers must have the required content understandings (e.g., knowledge of and about different SSI), the pedagogical knowledge needed for conducting a scientific inquiry about these issues as well as to internalize the attitudes needed to prepare students to take informed actions on SSI. This approach, therefore, challenges teachers to re-examine and to adapt their instructional practices which raises the need for teacher professional development programmes to prepare teachers to face these challenges. This need is all the more reinforced considering that SSI instruction is misrepresented in current professional development programmes (Hofstein et al. 2011).

The main objective of this chapter is to present and discuss the educational effectiveness of a teachers' professional development (TPD) model aimed at the development of science teachers' knowledge about SSIBL and about means to incorporate SSIBL into the teachers' practice. To address this objective, we subsequently present the design of a TPD model in two rounds.

8.3 Design of the TPD Model

The TPD course was developed in two rounds during two consecutive academic years (2015–2016, and 2016–2017). The National Center for High School Biology Teachers, located at the Weizmann Institute of Science and funded by the Israeli Ministry of Education, published the TPD course along with other professional development courses offered to biology teachers during each academic year. Twenty-two teachers from all over the country responded and participated in each of the two TPD courses (12 females, 10 males). During each round, the TPD course ran for 30 hours and included four face-to-face full-day meetings and one synchronous on-line meeting. The meetings took place during school holidays and were spread throughout the academic year (December to April), allowing the teachers time to implement projects in their classrooms between the third and the fourth meetings.

In the first round, the course was composed of two parts: (i) SSIBL implicit, and (ii) SSIBL explicit (Table 8.1). The first part of the course was SSIBL implicit and included two phases: Orientation and Experimentation. We began the TPD (the Orientation phase) with lectures from experts on complex social issues. We collaborated with one of the high schools in a central city in the middle of our country in order to exemplify a SSIBL-like project that runs in this school. In this project the cyanobacteria Arthrospira (Spirulina) is suggested as one solution to end world hunger. In the second day of the TPD (the Experimentation phase) the participants

Table 8.1 Outline of the two consecutive rounds of the TPD

	1st meeting Orientation	**2nd meeting Experimentation**	**3rd meeting Conceptualization**	**4th meeting Reflection**
	1st round of TPD course			
Activities	Lectures about current SSI	Lectures about current SSI	Introduction to the SSIBL framework	Teachers' presentations
	Controversy mapping	Inquiry activity	Examples of SSIBL in practice	Bridging between science and industry
	Exposure to student led SSIBL project	Discussion	Discussion	Discussion
	2nd round of TPD course			
	Lectures about current SSI	Examples of SSIBL in practice by graduate teachers	Examples of SSIBL in practice, planning for final projects	Controversy mapping
	Simulated inquiry activity	Conceptualization of core concepts	Assessment	Teachers' presentations
	Introduction to the SSIBL framework	Discussion	Discussion	Discussion
Epistemological explicitness			Implicit	Explicit

Grey backgrounds represent explicit presentation of SSIBL, while white backgrounds represent implicit presentation of SSIBL, in various parts of the TPD experienced a one-day IBL activity, which allowed them to perform experiments in the context of the high-school students' ongoing SSI spirulina project. We led lab experiments specifically intended to explore the optimal conditions for the spirulina's growth and protein content. The second part of the course was SSIBL explicit and included two phases: Conceptual and Reflection. On the third day of the TPD (the Conceptual phase) the participants were introduced to the practical and theoretical elements of SSIBL. This event marks the first time the teachers were provided with explicit details in relation to the theoretical thinking behind the PARRISE project. By the end of the third meeting, teachers were asked to prepare their own SSIBL projects to be implemented in their classrooms. On the fourth day (the Reflection phase), the teachers presented their planned projects, with subsequent reflection and discussion sessions.

In the second round, the TPD course was composed of the same four phases, namely Orientation, Experimentation, Conceptualisation and Reflection, but each part was modified according to the experience gathered in the previous year, and an Evaluation phase was added with the aim of connecting the SSIBL idea to the school curricula (Table 8.1). During the Orientation phase, expert lectures were delivered and the inspiring spirulina project was introduced as an example for an SSI project. However, this time the spirulina project was introduced as a 'dry' laboratory aimed to explore the optimal conditions for the spirulina's growth and for obtaining optimum protein content through minds-on instead of the hands-on experiences, used in the previous round. This day ended with discussions of the SSIBL idea.

During the second day of the TPD (the Experimentation phase), the teachers were exposed to several SSIBL examples. Two teachers who attended the TPD in the previous year each presented an example. These examples were of SSIBL projects the teachers had enacted in their classes during the previous year TPD (one of these teachers is presented as one of the case studies below). This day ended with discussions of the SSIBL theory, thus the Conceptual Phase started by the end of the second day. The teachers were then asked to plan potential opportunities for SSIBL projects that may be incorporated into the high school biology or environmental sciences curricula and to start designing projects that were in line with the SSIBL idea to be implemented in their schools. During the third day of the TPD, the conceptualisation phase continued with introducing the teachers to the ENGAGE (Equipping the Next Generation for Active Engagement in Science) project (ENGAGE 2014), as an example of another project in which social and scientific issues are combined including the evaluation methods used in this project. To complement the Evaluation Phase, which was added in this round of the TPD, we collaborated with the national supervisors for biology and environmental sciences education from the Ministry of Education and introduced a new initiative to examine students' abilities to answer 'OMER' (which stands for the terms Values, Involvement and Relevancy in Hebrew) questions into the national matriculation examinations. Thus, promoting the linkage and ensuring an alignment with the school curriculum. During the fourth day, the Reflection phase, the teachers learned from each other about their plans to execute SSIBL projects. The participating teachers presented their projects, shared their experiences, brought evidences from class, reflected on them, and proposed constructive recommendations for integrating SSIBL-projects in various high-school settings.

8.4 Overview of the Research Approach

In order to study the educational effectiveness of the TPD model presented above for the development of science teachers' knowledge about SSIBL, we conducted a qualitative study of two teachers' experiences, one who had participated in the first round of TPD and the other who had participated in the second round. As described above, the TPD was structured to educate and support the implementation of SSIBL in classrooms. Accordingly, in the second round, a few teachers participating in the first round were invited to present the implementation of SSIBL in their classrooms, thus making the SSIBL approach more concrete to the participants of the second round.

The results are discussed in relation to the following question: What can the experiences of teachers tell us about the educational effectiveness of a TPD model aimed at promoting the implementation of SSIBL in science classrooms?

8.5 Method

8.5.1 Participants and Data Sources

Experienced in-service high-school science (biology and environmental sciences) teachers participated in the TPD (30 h, 4 days, n = 22 in the first round and n = 12 in the second round). Here we focus on two case studies: (i) a teacher who participated in the first round of the TPD; and (ii) a teacher who participated in the second round. These two teachers were selected since in certain aspects they shared the same concerns in regards to teaching science, but their experiences from the course and the methods of implementing the SSIBL pedagogy in their classrooms were different. In addition, these two teachers were relatively more involved in the course discussions.

Data sources were in-depth interviews with the two teachers, their written projects, their reflections on the TPD, and their oral presentations. Teacher interviews, and TPD observation were audiotaped and later transcribed, analysed, and interpreted. The use of multiple data sources allowed for triangulation of data and were used as a strategy for the validation of results.

The next section presents the cases of the two teachers, David and Ruth. David participated in the first round of the course and Ruth participated in the second round.

8.6 Results

8.6.1 Case Study 1: David

David has been a biology teacher for 7 years. He teaches students aged 13–18 at a school that emphasises the importance of democracy as part of the fundamental values by which it operates. There are approximately 480 students in the school, which is located in a kibbutz-agricultural environment that specialises in fish farming by intensive cultivation. In addition, there is an ecological site near the school that is used by teachers and students for various educational purposes.

David was a prominent participant in the first round of the TPD course. It was evident that, regardless of his participation in the course, he believes certain aspects of civic education are linked with science education. He was one of the teachers who were invited to the second round of the TPD to present the SSIBL project he chose to implement in his school during the first round of the TPD. He is currently working to promote social relationships that revolve around topical issues of science and society, between schools serving different populations in Israel.

David sees himself as an unorthodox teacher. He describes an educational climate in which a pedagogical change is required, but is also difficult to attain. As a young teacher his pedagogical approach was mostly traditional, focusing on content from textbooks and 'cookbook' laboratory activities. Over time, this method of teaching became difficult for him, therefore, of his own initiative, his lesson plans were gradually directed towards the ecological site nearby his school. This method of teaching received the support of the school principal, which encouraged David to attempt to disseminate this method among the rest of the teaching staff and the subject supervisor. However, these attempts were unsuccessful.

The attempts to receive collegial legitimacy for his innovative teaching methods were part of his motivation for participation in the TPD course, which he describes as related to a 'pedagogy of change'. During the course, David met people of various professions as well as other teachers who possess the same interests and beliefs concerning science education, and his need to belong to a pedagogical milieu and to receive legitimacy for his teaching methods, was met.

> In this course you meet people who are not only people of science, but also people who share the same moral values. People who have interesting projects and stories. You meet all sorts of people in this course. So, I would recommend it, even if it is just for the moral value of it. It is more than focusing on technicalities of schooling, you learn something else. (David, interview)

For David, the integration of science education with moral and civic virtue seemed natural. His motivation to participate in the course and to conduct socio-scientific inquiry with his students is consistent, in part, with how he sees himself as a teacher and with his own personal vision regarding 'good' science education.

Receiving legitimacy in the TPD course setting and the new knowledge he acquired allowed him to deepen and expand his lesson plans. For his final project,

David presented a plan for a socio-scientific inquiry project with his students that centred on aquaponics. The project aligned with the school's local community and lived experiences as some of the agricultural efforts in the local environment are dedicated to intensive fish farming. The interaction between the students and their surrounding community, and conducting a scientific inquiry project that would benefit this community were important elements in David's project. In his presentation to teachers participating in the second round TPD, he made sure to establish the importance of fish farming in terms of human consumption and the need for sustainable food sources, presented the environmental damage that this type of agriculture system – from the pollution of nearby rivers to the wellbeing of the cultivated fish. His research project was thus aimed at finding a solution that would minimise these damages while still benefiting from the Kibbutz's economy and social fabric.

> When I work with the students we try to think what solution best serves us. The question is which solution can combine society, environment and economics? Just as I want this solution to be sustainable – it must meet these three terms. It must also be economical so that it would interest the fish farmers [...] We must create some interest (with the fish farmers), some shared interest for working together. (David, first round of the TPD course)

In their research project, David and his students tested the efficacy of a biological filter that purifies the water of harmful substances. Moreover, as a by-product, the process of pumping and purifying the water supports the growth of fruits and vegetables on a bed of purified water in what David described as 'green' hydroponic agriculture. David emphasised the importance of this type of agriculture, as it is not only sustainable, but could also provide the school's community with the option of purchasing organic vegetables at a relatively low price. That is, the solution that David and his students developed for the issue of intensive fish farming takes into consideration the economic interests of the fish farmers and the community's needs on a wider scale.

> The pool that creates the food, it can potentially produce 2,500 lettuce plants per month and we want this food to go somewhere for a nominal price, a place that needs this food. To our school, to schools with a population of weaker socio-economic backgrounds, or a population with disabilities, and some can go to our dining hall for a very very low price. (David, interview).

Although his inquiry project was operationally complex and required constant attention and maintenance by himself and his students, David did not view this as the project's weak point, but as a challenge for his students to face and solve – when he was asked about the option of commercial distribution, he said that 'as the years progress I strive to advance (my students) even further'. Much to his satisfaction, David's students were invited to give a workshop on their research to teachers and other professionals from other schools. This invitation, together with the fact that David was invited to present in the second round of the TPD course, further established the legitimacy of David's actions as a teacher and enabled the dissemination of his pedagogical ideas.

Although, at the time of writing, he is currently on sabbatical for a year, David is already planning additional far-reaching research activities that will connect with his research project in the future and he also plans to incorporate the mathematics teaching staff into these efforts.

> We will also study the effects of growing plants on a bed of rainwater and not just with drinking water. We also have this mathematical element of how to calculate [...] We have a rooftop that collects the rainwater, we have a certain amount of storage tanks, a certain amount of containers that we can keep, and a certain amount of trees and plants that need watering. How do we configure this as a system? How do we create a mathematical model that can tell us according to our roof's surface area and according to the average amount of rain in our area per year, how much storage we require? And how many trees and plants this could serve? (David, interview)

In addition to expanding the breadth and depth of his research project, David intends to continue to relate his students' scientific knowledge and activities to their own lives, their world, and the school's community. For David, this connection is viewed as both the means and the aim of scientific education and he states that in the upcoming year he plans to emphasise this connection.

> This is the point of taking a scientific issue and connecting it to the reality of our lives and to some basic needs that we possess. It's easy to lean onto some instructional routine that is familiar. Most of the times, my students don't care for the same things that I care about. They aren't interested in environmental issues. The challenge here is to keep an open mind. Not say 'Okay, I'm offering you these and that issues to work on', but to actually try (and connect with the students' interests), and it's challenging. You need to have the right tools to do this, how to start a conversation, how to focus the conversation around students' interests, and then see how you can connect their interests to some scientific research. Remember that our goal here is to involve the students with science and with social issues. This involvement doesn't simply happen by asking them on these issues in tests, but by providing them with something fundamental that will give them a sense of accomplishment and belief in their abilities. (David, interview)

In this excerpt, David summarized his view for appropriate science learning and instruction. For him, tapping into students' interests was a strategy to not only promote their civic participation but also to prevent their possible alienation from science and scientific practices. To achieve this aim, David was willing to put in the additional effort and to relate their experiences from outside of the school to their science learning experiences and the design of his inquiry activity was structured around that notion.

8.6.2 Case Study 2: Ruth

Ruth is a teacher with 25 years' teaching experience, who has been teaching biology for 16 years. She teaches young people aged 13–18 at a boarding school that serves approximately 800 students. The school emphasises an atmosphere of tolerance, sharing and openness and is located in the central area of Israel in an urban environment. The goal of the school, according to its credo, is to cultivate alumni who will be contributing and productive citizens, who will march society forward. For this reason, the school emphasises independence, maturity and initiative taking as core values. They emphasise respect for the law and democracy, and make sure to maintain human dignity.

Despite being a veteran teacher, Ruth participates in TPD courses since she believes it is important to be up-to-date on subject matter knowledge. Ruth states that she possesses a great deal of professional responsibility, which motivated her to be updated, advance her skills and sign up for TPD courses. Before the course she did not consider incorporating socio-scientific aspects into her science instruction. She came to the TPD course due recent educational policy demands, which required an emphasis on values, students' engagement with both the subject knowledge and civil life, and the relevancy of instructional content to the lives of students (the 'OMER' questions mentioned above). Unlike past years, students are planned to be tested and evaluated on these elements in future matriculation exams, which therefore support the implementation of SSI instruction in science classrooms. In a different manner to David, Ruth's motive for participation in the TPD course was therefore external and composed of changes in the policy and guidelines of the Ministry of Education.

Though she stated her satisfaction with the TPD saying that it represented 'a new point of view', the concept of SSIBL did not appeal to her. In theory, she supports a pedagogical approach that integrates inquiry with SSI, but does not envision it in practice in the educational field with large groups of students, as in her class. It is apparent that Ruth's instructional approach is mostly directed towards improving students' test scores on the matriculation exams.

> [For example], in the Bio-inquiry, I mainly focus on "let's do an experiment, let's write an essay". That, in itself, is not easy at all with the students, and this year I have 38 students who are divided into 16 small groups. We have to be focused, so I wasn't looking to add on something extra to this. [I only wanted to] focus on the familiar and known content and to finish this task. (Ruth, interview)

Nonetheless, she did find ways to combine materials from the TPD course in her activities at school, but she stated that she will implement SSIBL in the future only if it will be mandated by the subject superintendent or by the Ministry of Education. Though not supportive of SSIBL in practice, her participation in the TPD did allow her to cope with new policy demands to implement SSI in science instruction.

> [The new policy demands] *appealed to me and I implement it in my classroom. Also, it's easy for me to think of questions (that suit that new policy), I can improvise them very quickly.* (Ruth, interview)

Ruth said that she added discussions of SSI to her lessons on several occasions. She mentioned an example from a lesson in which she and her students discussed the question of using pig heart valves transplants for people with heart conditions in life-threatening situations.

For her final project in the TPD course, Ruth presented a plan for a socio-scientific project with her students, concerning the support and care of children with heart conditions in under-developed countries, within the setting of the Save a Child's Heart NGO. This voluntary organisation conducts an international humanitarian project, which locates children from developing countries who require life-saving heart surgery, and provides them with medical care and follow-up. For a whole year, Ruth and her students supported children with heart conditions from this organisation.

Ruth believes that the discussion of SSI enhances the students' motivation to learn science because it engages them with issues that are relevant to their lives and also develops their critical thinking. However, she had reservations regarding the intellectual level that is required of students to effectively participate in this activity. In her opinion, although it is important to incorporate this activity, it is only suitable for the high-level students in different age groups.

> It doesn't really suit students who are struggling with the basics, right? It is more suitable for the stronger kids in both junior high school and high school. (Ruth, interview)

Despite these reservations, Ruth made sure to distribute the knowledge she acquired in the TPD, and shared lesson plans and activities she conducted pursuant to the course with the community of teachers in her area.

8.7 Conclusions and Discussion

The two case studies presented above tell the stories of two high school biology teachers who experienced the SSIBL TPD. One of them, the less experienced teacher – David – had previous experience as well as an internal drive to incorporate SSI into his teaching prior to the TPD course. The integration of science education with moral and civic virtue seem natural for him. Indeed, during the course David implemented an ambitious project with his students. In this project his students experienced scientific inquiry which was tightly linked to their surrounding community, namely David's students had a purely SSIBL experience. In contrast, the other teacher, the more experienced teacher – Ruth – had no prior experience in incorporating SSI into her teaching prior to participating in the course. Ruth also had no internal drive to do so. Her motivation for participation in the course was external, as she herself stated that changes in the policy and guidelines of the Ministry of Education towards the matriculation examinations might enforce her to incorporate SSI into her teaching in the near future. She also claimed that the concept of SSIBL did not appeal to her and that she will implement SSIBL only if the Ministry of Education will mandate it. Accordingly, during the course Ruth carried out a purely SSI project with her students, namely supporting and caring for children with heart problems from under-developed countries. Taken together, these two case studies point to an advance in both teachers' practice as both of them incorporated new pedagogical approaches. But, it seems that the course enabled the 'SSI experienced' teacher to move towards implementing SSIBL while it enabled the 'non-SSI experienced' teacher to implement SSI, but not to implement SSIBL. Since the guidelines from the Ministry of Education do not mention SSIBL as part of the high school biology program, this conclusion should be taken with caution, as these guidelines seem to influence one of the teachers (Ruth) and not the other (David).

It was previously reported that when the SSI approach is adopted in schools, the instruction is mostly constrained to a presentation of the social dilemma with no attempt to promote students' meaningful participatory engagement or action (Bencze et al. 2012). This does not seem to be the case here. Since both teachers' approaches led students to engage with action, either in their attempt to grow fruits and vegetables using the purified water they prepared in the course of their project in David's case, or by supporting children with heart problems in Ruth's case. We assume that the practical aspect of the course, with the emphasis on implementing projects in schools and reporting on them, made the implementation of SSIBL practical to the participating teachers. Thus making SSI instruction practical and overcoming the previously reported claim that SSI is misrepresented in professional development programmes (Hofstein et al. 2011).

The two teachers presented here differed in several aspects including their motivation for participation in the TPD course, their prior teaching experience and their experience in integrating SSI into their teaching. As a result, they reached different levels of implementation of the SSIBL idea during the course. In future, in such TPD courses, the following aspects should be taken into account: (i) In the Orientation phase, an emphasis should be given to connecting the topics discussed to the national curricula (if they exist) and to the matriculation examinations, taking into account both content knowledge as well as the ethical aspects that are included in the syllabi; (ii) In the Experimentation phase, provide more concrete examples of SSIBL projects that were carried out in schools by graduates of the previous SSIBL TPD courses, as well as bring wet lab experiments that can be carried out in the school laboratory (instead of the 'dry' lab experience described here for the second round TPD); (iii) In the Conceptualisation phase, express the SSIBL components explicitly and make sure the teachers will use them in their projects in schools; (iv) In the Evaluation phase, provide in-depth opportunities to experience and to develop questions for the matriculation examination part that is focused on SSI issues (the 'OMER' questions part). A discussion as to how to evaluate the students' project should be added to the course as well; (v) In the Reflection phase, the participating teachers should present the projects they carried out in their schools. As mentioned above, we found this part to be one of the most important aspects in the course as, in addition to the importance of reflection to the teachers' practice, it also enables all other teachers to hear about practical ideas that they may be able to implement in future years.

Clearly, new approaches to science education and science learning pose critical challenges for teachers, especially ambitious and time-consuming approaches such as SSIBL. The TPD may have provided both David and Ruth with considerable knowledge and tools for effective implementation of SSIBL in their classrooms, as well as confidence in their abilities as teachers of SSI, however, the findings reported here are not an example of instant transformation. Though their experiences as TPD practitioners diverged in motivation and ways of classroom implementation, both David and Ruth felt that the legitimisation of SSIBL based instruction would ease the transition from their current classroom practices to be more in line with SSIBL. While David was seeking his own community of practice where this approach is perceived as integral to science education, Ruth legitimised her actions by connecting them to external national policy mandates. Therefore, even though both teachers showed implementation efforts and signs of commitment towards SSIBL, the lack of such perceived legitimacy might impede effective implementation. For the same reasons, sharing the products of their endeavours as TPD practitioners and hearing other teachers' stories of implementation was considered an important aspect of the TPD – it granted a much-needed legitimacy for teachers' actions and ways of teaching. However, in light of previous studies (Bencze et al. 2012; Bencze and Sperling 2012; Zeidler et al. 2005), it is likely that efforts promoted by these teachers will face some difficulties to achieve legitimacy in school systems. It can therefore be inferred that SSIBL will not be the norm in science classrooms without being first legitimized by school systems. Additional legitimising agents may include a coherent and focused representation in national curriculum and matriculation exams as well as the availability of SSIBL focused teaching materials.

References

Aikenhead, G. S. (2005). Research into STS science education. *Educación Química, 16*(3), 384–397.

Barton, A. C., & Tan, E. (2010). "It changed our lives": Activism, science, and greening the community. *Canadian Journal of Science, Mathematics and Technology Education, 10*(3), 207–222.

Bencze, L., & Carter, L. (2011). Globalizing students acting for the common good. *Journal of Research in Science Teaching, 48*(6), 648–669.

Bencze, L., & Sperling, E. R. (2012). Student teachers as advocates for student-led research-informed socioscientific activism. *Canadian Journal of Science, Mathematics and Technology Education, 12*(1), 62–85.

Bencze, L., Sperling, E., & Carter, L. (2012). Students' research-informed socio-scientific activism: Re/visions for a sustainable future. *Research in Science Education, 42*(1), 129–148.

Berkowitz, M. W., & Simmons, P. E. (2003). Integrating science education and character education. In *The role of moral reasoning on socioscientific issues and discourse in science education* (pp. 117–138). Dordrecht: Springer.

Dos Santos, W. L. (2009). Scientific literacy: A Freirean perspective as a radical view of humanistic science education. *Science Education, 93*(2), 361–382.

ENGAGE. (2014). *Engage – Equipping the next generation to participate in scientific issues.* Retrieved from http://www.engagingscience.eu/en

Hodson, D. (2003). Time for action: Science education for an alternative future. *International Journal of Science Education, 25*(6), 645–670.

Hofstein, A., Eilks, I., & Bybee, R. (2011). Societal issues and their importance for contemporary science education—A pedagogical justification and the state-of-the-art in Israel, Germany, and the USA. *International Journal of Science and Mathematics Education, 9*(6), 1459–1483.

Israeli Ministry of Education. (1992). *Tomorrow 98: Report of the superior committee on science mathematics and technology education in Israel.* Jerusalem: State of Israel Ministry of Education Curriculum Center.

Kolstø, S. D. (2001). Scientific literacy for citizenship: Tools for dealing with the science dimension of controversial socioscientific issues. *Science Education, 85*(3), 291–310.

Lee, H., Chang, H., Choi, K., Kim, S.-W., & Zeidler, D. L. (2012). Developing character and values for global citizens: Analysis of pre-service science teachers' moral reasoning on socioscientific issues. *International Journal of Science Education, 34*(6), 925–953.

Levinson, R. (2010). Science education and democratic participation: An uneasy congruence? *Studies in Science Education, 46*(1), 69–119.
Levinson, R., & PARRISE-Consortium. (2014). *The SSIBL framework*. Utrecht/London: D1 2 PARRISE, co-funded by the European Commission under the 7th Framework Programme/ Institute of Education. Retrieved from http://www.parrise.eu/
Levinson, R., & PARRISE-Consortium. (2017). Socio-scientific inquiry-based learning: Taking off from STEPWISE. In *Science and technology education promoting wellbeing for individuals, societies and environments* (pp. 477–502). Cham: Springer.
Roberts, D. A. (2013). Scientific literacy/science literacy. In *Handbook of research on science education* (pp. 743–794). New York: Routledge.
Roth, W. M., & Lee, S. (2004). Science education as/for participation in the community. *Science Education, 88*(2), 263–291.
Sadler, T. D., & Zeidler, D. L. (2009). Scientific literacy, PISA, and socioscientific discourse: Assessment for progressive aims of science education. *Journal of Research in Science Teaching, 46*(8), 909–921.
Sadler, T. D., Barab, S. A., & Scott, B. (2007). What do students gain by engaging in socioscientific inquiry? *Research in Science Education, 37*(4), 371–391.
Walker, K. A., & Zeidler, D. L. (2007). Promoting discourse about socioscientific issues through scaffolded inquiry. *International Journal of Science Education, 29*(11), 1387–1410.
Zafrani, E., & Yarden, A. (2017). Becoming a science activist: A case study of students' engagement in a socioscientific project. *Sisyphus – Journal of Education, 5*(3), 44–67.
Zeidler, D. L., Sadler, T. D., Simmons, M. L., & Howes, E. V. (2005). Beyond STS: A research-based framework for socioscientific issues education. *Science Education, 89*(3), 357–377.

Dr. Rachel Cohen is the coordinator of the Rothschild-Weizmann Program for Excellence in Science Education and the Teaching Certificate Program at the Feinberg Graduate School of the Weizmann Institute of Science. She held various positions in the Israeli Ministry of Education and worked extensively with pre-service as well as in-service junior high schools science teachers and with senior high school biology teachers on implementing inquiry into their teaching. Her research focused on expression of the assimilation of a change in teaching the living cell as a "longitudinal axis" (macro-micro) in junior high school and the various aspects of its curriculum implementation. In light of her research findings, an intervention program (in-service training course) was developed and conducted, dealing with teachers' pedagogical-content knowledge (PCK), to improve the expression of the new approach.

Eran Zafrani is a PhD student at the Department of Science Teaching at the Weizmann Institute of Science. His M.Sc. thesis dealt with social and cultural aspects of science learning and teaching. Specifically, he examined the ways in which high-school students participate in a socioscientific project using identity research as an analytic lens. His current research aims to understand more about how field-wide institutional factors support or constrain science teachers' from implementing instructional practices that include dialogic argumentation in their classrooms.

Professor Anat Yarden is Head of the Department of Science Teaching at the Weizmann Institute of Science, and Head of the Biology Group at this Department. The primary theme in all of her academic activities has been the attempt to adapt practices employed by scientists, to the processes by which students and teachers accumulate and advance their knowledge within the discipline of biology. Towards this end, her group pioneered the adaption of primary scientific literature for the teaching and learning of biology in high schools.

Personal page, Weizmann Institute of Science, Israel. http://www.weizmann.ac.il/ScienceTeaching/Yarden/home

Chapter 9
Getting Ready to Work with Socio-Scientific Issues in the Classroom: A Study with Argentine Teachers

Melina Furman, Inés Taylor, Mariana Luzuriaga, and María Eugenia Podestá

9.1 Introduction

Many voices argue that there is a disconnect between the way science is traditionally taught in many school classrooms and the complex and creative approaches required to solve socio-scientific problems in real life. Research shows that in the Latin American region in general and Argentina, the context of this study, in particular, typical science lessons involve students spending most of their time memorizing facts and definitions, or 'proving' pre-existing relationships through demonstrative practicals (Valverde and Näslund-Hadley 2011; Furman 2018; Furman et al. 2018). This approach dominates despite most science curricula in the region stating their intention of using science education as an opportunity to develop conscientious citizens capable of the critical thinking skills needed to navigate the twenty-first century (Miller 2000).

One way of promoting those critical thinking skills in students within science education is through the incorporation of socio-scientific issues (SSI). Approaching science education from an SSI standpoint allows students to start to view problems for the complex, messy and multi-faceted challenges they truly are (and therefore more closely resemble the challenges faced by actual scientists and critical citizens). For this chapter, we define SSI as those which invite students to think in more complex ways, requiring them to discuss, debate, negotiate and understand connections to resolve conflicts (Zeidler and Nichols 2009). Through SSI, students can be

M. Furman (✉)
Universidad de San Andrés, Victoria, Argentina

CONICET, Buenos Aires, Argentina
e-mail: mfurman@udesa.edu.ar

I. Taylor · M. Luzuriaga · M. E. Podestá
Universidad de San Andrés, Victoria, Argentina
e-mail: itaylor@udesa.edu.ar; mluzuriaga@udesa.edu.ar; mepodesta@udesa.edu.ar

M. Evagorou et al. (eds.), *Science Teacher Education for Responsible Citizenship*, Contemporary Trends and Issues in Science Education 52,
https://doi.org/10.1007/978-3-030-40229-7_9

encouraged to develop new thinking habits, such as skepticism, accepting ambiguity and open-mindedness, whilst searching for answers to complex problems (Zeidler et al. 2005). Importantly, using a SSI standpoint invites students to contemplate, amongst other things, their moral and ethical opinions about scientific topics through social interaction and discourse (Lee et al. 2013; Zeidler and Keefer 2003). SSI approaches also have the potential to achieve other goals of science education more generally (Sadler 2011), such as encouraging students to make data -or evidence-based decisions, whilst also evaluating the quality of available information. As such, SSI can be viewed as an approach to science education that provides students with the chance to develop the skills required to be full members of society (Díaz-Moreno and Jiménez-Liso 2013; Sadler et al. 2007). This chapter explores how Argentine teachers responded to a first introduction to teaching with socio-scientific issues. Following an initial workshop held in the province of Buenos Aires, we explored teachers´ perceptions of the challenges and opportunities associated with incorporating these strategies into their regular lessons. A case study of three teachers is then explored in more detail as they introduce SSI approaches to their classrooms.

9.2 Background

As teachers move away from more fact-based and 'black or white' science teaching, as is the norm in Latin America, there is a need for different pedagogical tools to help students address the sophistication and uncertainty involved in dealing with complex issues. For this change to happen, teachers need to use various pedagogical approaches and lines of questioning to expose students to different opinions and viewpoints in the classroom, particularly regarding those topics which most impact and engage young people in political debate (Gray and Bryce 2006; Nielsen and Evans 2015).

However, although research has shown that working with socio-scientific issues in the classroom is an effective way of promoting aspects of scientific literacy and students' understanding of the nature of science (España Ramos and Prieto Ruz 2010; Wongsri and Nuangchalerm 2010), studies also indicate that teachers find this approach challenging (Pitiporntapin et al. 2016). In particular, teachers have been found to be reluctant to teach using SSI approaches due to concerns about their abilities, time constraints and lack of support materials (Pitiporntapin and Topcu 2016). This is particularly the case in Argentina, where teachers tend to favor an encyclopedic approach to science, focusing on learning facts and definitions and seldom promoting higher-order thinking skills in authentic contexts. For example, a recent study by Furman et al. (2018) found that teachers spend 80% of their science lessons undertaking activities that only promote lower-order thinking skills, as opposed to activities which foment creativity, evaluation or synthesis of problems.

One of the reasons that this type of teaching might prevail may be that, despite Argentine curriculum documents mentioning the value of incorporating SSI, teacher education programs in the country (either pre- or in-service) do not often include this approach with sufficient depth or prepare teachers for engaging students in the debate of socio-scientific dilemmas (Labate 2007). As research has shown, supporting teachers in reviewing and transforming their teaching practices is a constructive process which depends on where teachers are in their current practice, from which they must make sense of new pedagogical approaches in terms of their existing beliefs (Pugh et al. 2017). For teachers to be able to adapt or fully appropriate new pedagogical approaches or even teaching techniques, they first need to be introduced and initiated before moving along subsequent phases of teacher expertise (Dwyer et al. 1991).

Another challenge may be due to the views that Argentine teachers commonly hold regarding the nature of science. As is the case with many countries, this view of science, which differs from the complex, humanist view suggested by the SSI approach, presents science as a static body of knowledge which needs to be memorized, rather than a complex, adaptable agglomeration of ideas (Pujalte et al. 2015). Teachers' views on the nature of science are shown to influence their practices (Caga-anan and O'Toole 2015), and this impact is heightened by the fact that tertiary teacher training colleges (pre-service teaching) do little to challenge these views, allowing this conception of science to go unquestioned right until teachers find themselves at the front of a new classroom (Cofré et al. 2015).

Despite this state of affairs, Argentina also presents several context-specific opportunities with respect to introducing SSI in the classroom. As national examinations are anonymous and results are not publicly published, teachers and schools are free of the pressures and accountabilities related to preparing students for high-stake examinations. Also, although teachers have a prescribed national curriculum they are responsible for covering throughout the year, systematic teacher performance reviews are infrequent and not linked to pay, meaning that teachers have a remarkably high degree of autonomy in terms of what and how to teach. As a result, one could argue that this scenario opens the possibility of introducing new ways of understanding and teaching science, such as through SSI approaches.

Given the above state of affairs, in this study we aimed to understand how teachers started to incorporate elements of SSI approaches in their science lessons following a professional development program.[1] We devised an in-service professional development workshop that focused on teaching science creatively through the use of SSI. Subsequently we arranged a personalized follow-up which supported teachers in the introduction of elements of the workshop back to their classrooms. We were interested in understanding how teachers started to implement SSI approaches, including the challenges and opportunities they identified, as well as the effect that these approaches had on students´ perspectives of their engagement with science

[1]This was part of a wider initiative undertaken in collaboration with Donald Gray and Laura Colucci-Gray from the University of Aberdeen, generously financed by an International Partnership and Mobility grant from the British Academy (2013–2016). The joint project aimed to foment creative practices in science more generally, including a focus on SSI, in both Scotland and Argentina.

lessons. Lastly, we explored how teachers' use of SSI approaches changed and developed over a 2-year period. As such, our research questions were:

1. How do teachers incorporate SSI approaches in their own practices following a workshop? How are these changes sustained over time?
2. How do students respond to the introduction of socio-scientific issues? In what ways do their perceptions of their engagement with science lessons change?

9.3 Methods

9.3.1 Research Design Overview

To answer our research questions we divided the study into two parts. In the first part, an in-service teacher development program was conducted. All 50 participating teachers filled in a post-workshop survey, and were invited to volunteer for a follow-up study.

For the second part of the study, follow-up teachers worked collaboratively with researchers to plan and then implement lessons with an SSI focus. Following the implementation and observation of the lessons, teachers completed an in-depth semi-structured interview. Lastly, one year after the initial workshop, teachers were invited to participate in a final lesson observation and interview.

9.3.2 Part 1: In-Service Professional Development Workshop

For this study, two 3-h twilight workshops were designed during which teachers were introduced to the idea of SSI perspectives with an additional focus on using creativity to enrich science lessons. The sessions included concrete examples of teaching activities that could serve as good models for similar approaches in their lessons. Teachers were recruited via usual university outreach mechanisms, including mailing lists and social media posts. Teachers were from a variety of professional backgrounds, although predominantly from private institutions, ranging from several preschool and primary generalists looking to approach science teaching differently, to secondary school biology and chemistry teachers with specific subject matter interests.

9.3.2.1 Workshop Contents and Activities

The program was designed by the authors in collaboration with the two colleagues from the University of Aberdeen. When designing the overall course, several features of effective teacher training served as guiding principles. As stated by Gray and Bryce (2006, p.187):

> just as 'top-down' transmission of facts from teacher to student is inappropriate, so 'top-down' delivery of CPD from 'expert' to teacher is inappropriate without a concomitant commitment to explore the issues and provide opportunities for reflection and personal feedback on them, as well as exploration of appropriate pedagogical approaches to be used in the classroom.

As such, our CPD incorporated expert input, as well as individual and group activities, with opportunities to debate and listen to others from an active learning and reflective standpoint.

Teacher professional development research shows that for changes in teaching practice to occur and be mastered, a minimum of 14 hours of training is needed, as well as many instances for practice (Joyce and Showers 2002; Yoon et al. 2016). As such, these workshops were designed to be introductory in nature, rather than transformative in terms of teacher practice, and to explore teachers' perceptions of how SSI approaches might fit into their teaching more broadly.

The workshops were based around the topic of agricultural practices, which was framed as a complex socio-scientific issue, taking into account the way it connects to issues surrounding equity, land distribution, impact on the environment, conservation, health and community. The problems selected were also purposefully chosen to be 'messy', that is, those that involve multiple-causality and require the analysis of different perspectives in order to reach one of many possible solutions.

The first workshop session invited teachers to explore the relationship between soil and agricultural practices. This was also explicitly linked to 2015 (the year when the workshop was conducted) being the United Nations 'International Year of Soil'. By looking at soil and its intricate relationship with agricultural practices relating to food production, survival of communities and sustainability, teachers were invited to evaluate the context of different countries and the practices they chose, as well as consider how different agents and stakeholders value the diverse range of opportunities available to them through different agricultural practices.

Teachers began the workshop by considering the role of soil in our society. This topic was introduced through the use of a video that outlined how soil shapes and sustains many human activities, told from the perspective of the soil itself and ending with a plea to consider its importance. Teachers used further information to mind-map the intimate relationship between soil quality and human survival. Following this introduction, they were invited to complete a self-designed experiment in groups, for which they were given four soil samples of different constitutions (comprised of sandy, clay-based, and mixed soil samples) and asked to evaluate the permeability of each soil type to water using their own methods. With the results garnered from their own data, each group was asked to decide which of a choice of several crops they would recommend for growth in each soil type, taking into account the specific social, cultural, economic and environmental factors which had been outlined for each case.

Throughout the workshop (see Fig. 9.1), teachers were encouraged to view how this topic and approach, taken from a real world perspective with multiple standpoints on the problem of deciding which crops to plant where, invited different teaching and learning skills. For example, teachers had to use creativity and col-

Fig. 9.1 Images from the first workshop

Fig. 9.2 Images from the debate

laboration to design their 'measure of water permeability experiment', as they had not been given a method to follow, but rather a large choice of everyday equipment with which they had to design their experiment.

The second workshop session expanded upon the topic of soil and food production and was based around a debate regarding the use of organic versus industrial production techniques. For this session, teachers were split into groups and given a position and an identity from which to argue their point, along with background information on which to base their characters. For the side that was promoting the use of industrial production, teams were asked to embody and present the views of a biochemical engineer, a large-scale industrial farmer and a FAO (Food and Agriculture Organization) representative. For the side that was promoting organic production, teachers were asked to take the perspectives of an organic farmer, a parent and a bee. From each of these given positions, teachers had to prepare a 3-min exposition, followed by a question-and-answer session where the views, preferences and interests of their character were presented (see Fig. 9.2). At the end of the workshop, following a general debate, teachers had to place themselves in the role of legislators, and vote on whether to expand or retract the use of industrial practices in their local constituencies, having listened to the views of all the various characters. This workshop therefore centered on the themes of looking at problems from multiple perspectives, encouraging empathy and problem-solving, as well as working in teams to collect, analyze and present information.

Another theme that was maintained across both workshops was critical reflection. In both sessions, teachers were invited to reflect not only on their classroom practices, but also on their experiences as active learners in the workshops. As such, teachers were encouraged to think about what they learnt but also how they learnt it. Teachers were also asked to imagine how their students might react and respond to similar activities should they implement them in their schools.

9.3.2.2 Final Survey

At the end of the workshop, all the teachers completed an exit survey to explore the possibilities and challenges they anticipated with regards to introducing socio-scientific approaches to their current science pedagogies, and the contexts and factors that could shape their implementation.

9.3.3 Part 2: Follow-Up: Lesson Planning and Implementation

All teachers who participated in the workshops were asked if they were interested in participating in a follow-up study aimed at supporting teachers in the planning and implementation of workshop strategies regarding SSI approaches in their classrooms.

9.3.3.1 Participating Teachers

Ten teachers signed up to participate in the follow-up. For this study, we will focus on the three teachers who continued to work with us over the course of both years (see Table 9.1). Other teachers did not choose or were unable to continue the full

Table 9.1 Follow-up teacher characteristics

Teacher	Level	Education degree	Teaching context
Julia	Secondary	BSc Chemistry	Private school
			Chemistry, Physics and Physical Chemistry
Sofia	Secondary	BSc Biology	Private trilingual school
			Teaches Biology, Chemistry and "Health and Adolescence"
Clara	Kindergarten	BA in Kindergarten teaching	Private bilingual school
			Teaches 3-year-olds all subject areas

program for a variety of reasons, such as leaving teaching the following year, focusing on other school subjects (in the case of primary teachers) or having conflicting schedules.

9.3.3.2 Content and Activities of the Follow-Up Program

Teachers who participated in the follow-up worked with researchers to develop a series of lesson plans that involved socio-scientific issues. They worked together to analyze effective ways of undertaking the topics selected, and then to plan lessons for their respective age groups. In part, this follow-up was created as a way of accompanying teachers as they tried this novel approach for the first time, as 'implementation dips' can bring about frustration during the process of applying new approaches (Gulamhussein 2013), especially before seeing changes in student learning which may in turn convince teachers of their value (Guskey 2002). In this sense, having the researchers available to co-plan and support teachers during this initial phase encouraged teachers to plan ambitious lessons without feeling unsupported.

Once lesson plans were completed, all teachers were then observed when teaching their co-planned lessons, and given brief coaching feedback following implementation. During observations, researchers took notes and photographs, looking at student engagement and outcomes. In total, the follow-up program had a duration of over 20 h, above the minimum required to begin to see changes in teacher practice (Yoon et al. 2016). Table 9.2 below presents a synthesis of the methods of data collection used and the characteristics of the observed lessons in each case.

Table 9.2 Methods of data collection and characteristics of lessons observed by researchers per case

		Sofia	Julia	Clara
Grade		Secondary School	Secondary School	Kindergarten (3 year olds)
Subject		Biology	Chemistry	Science (interdisciplinary projects)
Year 1	Teacher interviews	1	1	1
	Class observations	2 (weekly, 60 min)	2 (weekly, 120 min)	2 (weekly, 30 min)
	Student interviews	1 focus group (4 students)	1 focus group (5 students)	1 focus group (3 students)
Year 2	Teacher interview	1	1	1
	Observation and subject	1 – Health and Adolescence	N/A	N/A
	Student interview	1 focus group (4 students)	N/A	N/A

9.3.3.3 Teacher Interviews

Following the secondary school lessons, teachers participated in a 45-min semi-structured interview, aimed at understanding their views about science and approaching science from a socio-scientific perspective, as well as the challenges and opportunities they found. In addition, we asked background questions to establish their professional biographies and work trajectories both within and beyond teaching. In order to gain a more tangible sense of the changes to their practice as a result of the workshops, we also asked and collected information regarding their previous lesson plans and approaches.

All interviews were transcribed verbatim and then analyzed by researchers. Data was analyzed through thematic analysis, and then coded according to the themes of motivational drivers, current science practice, beliefs on the purpose and nature of science education as well as professional experience and expectations. A mixed approach to coding was undertaken, with researchers searching for codes deductively (such as their views and beliefs on the purpose and nature of science, concerns and experience of starting new approaches and participating in professional development, and current teaching practices in science lessons), as well as including other themes that emerged through an inductive process of looking at the data. Interviews attempted to provide further insight into school characteristics, including opportunities for professional development, collaboration with colleagues, evaluation and assessment practices and support available.

One year later, all three teachers agreed to a second round of interviews. In addition, Sofia invited us to a further observation of a lesson. Both the interviews and observations followed the guidelines, topics and main questions of the previous year.

9.4 Results

9.4.1 How Do Teachers Incorporate SSI Approaches to Their Own Practice?

Teachers identified several aspects of the workshop that they believed could provide positive opportunities for their students, such as teaching in contextualized, authentic ways, promoting creativity and problem-solving, as well as increasing student engagement. However, one of the main perceived challenges was around time – both in terms of "imagining and planning" as well as implementing longer SSI-type lessons.

We then looked at how teachers took what they had learned in the initial workshop and examined the process by which they incorporated SSI approaches into their own practices, and how they evolved over time and persisted one year later. The three teachers who participated for the year-long follow-up came to the workshop for different reasons. Although all three mentioned wanting to increase student motivation, in the case of the secondary school teachers this was geared towards bringing the content alive in new ways, whereas Clara, the kindergarten teacher, sought more conceptual tools to be able to honor her students 'fascination with science' (which she personally struggled with).

As we will describe below, the way the teachers incorporated and sustained these approaches was profoundly shaped by their starting point regarding their content knowledge and their views on the nature of science learning.

9.4.2 Case Study 1: Clara – A Kindergarten Teacher

Clara was a kindergarten teacher at a private boys' school (her students were 3-years-old at the time of the study) who, by her own admission, regularly 'skived' science despite knowing that her students were very motivated and interested in learning about the natural world around them. As an early-years practitioner, she was already used to looking at contextualized problems (the Argentine early-years curriculum focuses on 'investigation of the natural, social and technological environment', rather than suggest particular or specific contents), but identified herself as being weak in science content knowledge.

She said she had left the workshop motivated, feeling like the experience had 'opened a window' to teach more and better science lessons. At the beginning of the follow-up process, Clara said she felt a little lost, both with the content and about how the unit would work with her children. As she noted: *'I was afraid because I was thinking "how will it go?", "will it interest them?", "will I be able to explain well?", "will I achieve what I want?", "how can I reach my learning objectives and be able to ask the kids good questions so that they can think, reason?'*

With the help of the researchers, she created a unit on the topic of soil and water discussed in the workshop. In the first lesson she introduced the idea of how water and the soil can be polluted and how to avoid this from happening, based on a story that involved the viewpoints of several human and animal characters that lived near a polluted water source. After listening to the story, children discussed what would be a good way to clean the 'dirty' water, which led the way to an experiment during which students first modeled polluting an existing water source (by throwing garbage into a large pail of water), and then working in small groups to filter the water using several different materials. This activity was similar in spirit to the activity teachers experienced in the workshop. Then, Clara worked with the physical education teacher to plan an interdisciplinary experiential activity, where students played 'catch' as either water molecules or pollutants – when 'waters' were tagged by 'pollutants' (who added a small velcro cloth to their clothes) it meant they then needed to go through a tunnel to get filtered to undo their being contaminated.

Throughout the unit there were several moments when Clara linked the science content with the SSI strategies suggested at the workshop, encouraging students to consider how it affected the world around them from different points of view.

Despite their young age, children discussed in their own terms who was responsible for looking after our water sources, how to avoid them getting more polluted, the consequences of having polluted water sources in terms of the dangers to the wildlife and human health, and how we can filter and improve water sources.

One of the interesting aspects of working with SSI approaches in kindergarten is that teachers are already used to working in contextual and embodied ways. In the case of Clara, she often reflected on how integrating the approach felt 'natural' and aligned with her usual practice and teaching mindset. Clara attributed her 'success' to the provision of detailed guidelines and materials at the workshop and the support given by the researchers with regards to how to best adapt these activities for kindergarten: *'I use your materials as examples of what to do. It's just better because it's already planned.'*

However, Clara also highlighted that she would not have felt confident enough to try these types of lessons on her own without the support of the researcher (*'I wouldn't manage to do it alone'*), especially on new science topics that went beyond those covered in the workshop. In fact, she mentioned having repeated the same unit in the second year but reported that she did not invent or develop further units or activities on new science topics.

9.4.3 Case Study 2: Julia – A Secondary School Chemistry and Physics Teacher

Julia was a secondary school Chemistry and Physics teacher who arrived at the workshop aiming to find new strategies to involve her students more during lessons. She was very confident in her content knowledge and wished to 'demystify' science, encouraging her students to feel comfortable with and enjoy learning it.

Over the follow-up process we noted that her first draft lesson plans implied that her lessons tended toward more traditional approaches, spending a large amount of time resolving formulaic exercises and memorizing content. This was confirmed by her students' comments from the subsequent interview, which described their role in the classroom as mostly passive: '*We don't usually do much, just sit around, listen, nothing*'.

For her follow-up unit Julia chose to work with her final year Chemistry class, where she had to teach the topic of colligative properties. As she described, she had taught the same topic in the previous year in a traditional manner: '*This was a topic I used to skim over. We'd discuss the properties theoretically, then look at the formulas and complete the exercises, followed by going to the lab. Classic format*'. This time around, she planned to introduce an overarching theme of 'how chemical and physical properties are present in real life situations'.

During her lessons, she was able to introduce new pedagogical strategies and resources, such as several group-work dynamics, videos, quick experiments and case studies, based around broad questions that aimed to connect the topic to real life situations such as 'should we throw salt on the ground when it's icy?' (encouraging students to analyze city policies in winter). Yet, a large part of her lessons were still based on traditional teacher exposition, and, in fact, her lessons were somewhat characterized by going over the same topics using different activities, rather than truly looking at a complex topic from multiple perspectives or encouraging reflection on the impact of science and technology advances on the environment and the community, which are the key characteristics of the SSI approach. In this case, it appeared that the activities planned with the SSI focus in mind were added as a complement or 'additional' aspect to the more central and fundamental learning goals.

Julia expressed enjoying teaching the follow-up lessons and, whilst stressing that she did not feel that these lessons had differed considerably from her normal practice, she did highlight the change she felt in her role as a teacher, which she perceived as having been '*removed from the center of knowledge, so that knowledge then gets created by everyone, which is also liberating for a teacher to be able to get out of the spotlight*' and incorporating more explicit links to everyday situations. With regards to how her students responded to the lessons, she felt that her students '*participated more, and thought more for themselves*'. This perception was echoed by several of her students, who also mentioned that they had 'copied out' much less in the new unit, and particularly enjoyed working in groups as this was not common practice.

Julia found the experience of planning and implementing the SSI approach to her lessons more challenging than anticipated (albeit with our support). In particular, one of her greatest concerns was with regards to the time required to plan and 'think' about these types of lessons, which coincides with the initial responses of the teachers in the workshop. As she said: '*I had to give myself space which I didn't have to be able to do this, it took time. I got a bit stressed*'. She added that she would value the institutions giving her more (paid) time to collaborate with colleagues and develop new strategies:

> Time is the scarcest resource, and in this case I had a lot of facilitators, I know. I mean, just resources, but I need time to go buy the rope, to get the materials. To think! Teachers need time to think. Time is the biggest obstacle.

In the second year, Julia commented on the changes she had sustained over the course of both years, and she mostly made reference to using several active learning strategies from the workshop (such as using a worksheet she had received, or using teaching stations and getting the students to move) to other science topics. However, as in the first year, the SSI approaches were not integral to the core of her teaching units.

Some of her comments from the second year echoed the views of other teachers who had participated in the workshop, who felt SSI approaches were not 'rigorous' enough when teaching specific chemical formulas and would 'water down' the science content. Julia felt that sometimes more active pedagogies interfere with the main science objectives: '*It's good to think and discuss, but how do we take that to the concrete formula they need to learn?*'. This comment seems to suggest that although she values the more participative approaches as a complement to traditional teaching, she does not appear to fully trust them since they are not aligned with her views of learning (that is, what content should be learned and how).

9.4.4 Case Study 3: Sofia – A Secondary School Biology Teacher

Sofia was also a secondary school teacher who had only recently entered teaching following a career as a research biologist. Perhaps due to her professional experience, her views on science teaching were considerably different to those held by Clara and Julia in that she intrinsically viewed science and science learning as a creative and social process, which she therefore tried to replicate in the classroom, showing students that science was 'real':

> As I always did research and I'm not a teacher [...] from that place I try to transmit that to them. I always tell loads of personal things about my own research, for them to realize that it's real, it's not something that we only stage for a science lesson artificially. I look for a lot of current information, I try to bring new things; newspaper articles, especially in health and environmental matters, we always work with new material and almost never with the books.

However, she recognized she lacked pedagogical strategies and generally relied on traditional activities without asking too much about the meaning of what she taught and about the relevance of what they were learning for the lives of the students. She signed up to the workshop as part of her wider goals to develop her teaching practice, but made specific reference to feeling she needed more tools and ideas to be able to 'reach' her students in different ways.

After the workshop, Sofia replicated every single one of the activities introduced with her classes (not as part of the 'official' follow-up, but on her own terms across different year groups).

For the follow-up, she chose to work with her most challenging group on DNA, which she said they were struggling with. She chose this class as she felt the group was struggling, identifying challenging behavior, weak academic skills and poor study habits as the main factor to explain this. Also, she felt that the topic itself was challenging and 'very conceptual', and somewhat harder to bring down to concrete simple examples, and would benefit from being learned in a more concrete and contextualized way. As she said:

> For me, my concern was the group. They are very good guys but they have quite bad study habits, they have a hard time understanding many things compared to other groups I have, they are quite disorganized. I was afraid they would disband or be lazy [...]. The other issue is that for me the subject is difficult, it is a conceptually difficult topic and super important. This helped me think about priorities, lower my anxiety and ask: 'Is it important that they understand this at the molecular level?' and just focus on the stuff that actually matters. Make them realize how important everything is and what is it for? That it's not something abstract, and that cells have to do with your life itself and your health and a lot of things.

The students had recently been learning about DNA and gene structure, and she used this as the content starting point to introduce the concept of genetically modified plants and animals to best suit human needs. The students watched an engaging TED talk on the topic of bioengineering and were then asked to get into groups of four. The activity reflected the key attributes of the SSI approach: examining a socio-scientific problem under multiple perspectives, using data to make decisions and reflecting on the impact of science and technology on human lives and the environment. The scene was set so that the students were asked to be an interdisciplinary team working for the government of a fictional developing country, whose characteristics were given on a factsheet (such as having a famine, low water supply, low immunization rates, to name but a few). Students were then invited to propose the creation of a genetically modified organism that could help solve some of the country's hardships. Students each had a role and a part they needed to explain (a scientist to explain the biology, a lawyer for ethics, a PR specialist to present the project in a public-friendly way, and graphic designer for the poster). Students presented in the second lesson, and then voted on which was most likely to effectively alleviate some of the problems the country was facing.

In line with our observations, Sofia perceived that the lessons had been successful, with all groups presenting a reasonable solution to the situation, and taking into account multiple viewpoints as well as the pros and cons of each suggestion made. Sofia felt that she had had more fun and enjoyed teaching more than in her previous lessons, as well as feeling that her students had been more engaged and better behaved during the lesson. Particularly she felt that assigning the student roles had worked well, as, although it required more careful planning and effort from the teacher beforehand, it then meant that the students were more autonomous and made the lesson 'less tiring' to teach.

In the following year, we could see how Sofia took the SSI approach a step further. She developed a new unit based on strategies introduced in the workshop for her Health and Adolescence subject. For that unit, she used the same debate format on the topic of abortion, a highly controversial topic for Argentine society, which is rarely included in secondary school lessons) from the point of view of several different roles (parents, religious leaders, a baby, a medical professional and a legislator). This unit was very socio-scientific in nature, with students fully embodying the different viewpoints and sustaining their roles based on evidence-based arguments. She also used the debate format with older students in her Environment, Development and Society classes on the topic of open mining:

> And after the format of the debate that you saw today, I used it last year with the same theme of abortion and I used it in Environment, Development and Society in the 6th grade. That worked really well as they're older. I used it for open pit mining, and we had the different characters; one was the governor of San Juan, another was the CEO of a mining company, another was a worker who had got a job thanks to mining. And then there was a Greenpeace representative, a vicuña [NB: an Argentine llama] – I remembered the bee and I liked that about the animal, and a native inhabitant who had his goats, for whom the changes in the area were a problem.

Sofia reflected on how, more than a year after the workshop and the initial follow-up process, she felt more open and willing to incorporate SSI approaches in her lessons, and that she looked for opportunities to be able to incorporate it in different topics. She shared how, at the beginning of the school year, she had looked over the curriculum searching for opportunities within which to incorporate the approaches she had learnt:

> At the beginning of the year I sat down with the syllabus and everything, I rethought about how to do many things in a different way from what I had been doing so far with the aim of incorporating these type of tools and approach to new topics.

Although she had full institutional support to teach in different ways with new strategies, similarly to Julia, Sofia also identified a lack of time in which to teach using these methods, particularly for the younger students:

> The truth is that we have to comply with the curriculum and I need more Biology hours for each class (which is impossible) or I need to convince myself that I don't care about the curriculum and it really is an internal decision to sacrifice content, to be able to do more meaningful things because they take more time. And more personal time too, because it's a lot of my own time and work. But if something works you can use it again, it's an initial effort but it's an investment.

9.5 Conclusions: What Does It all Mean for Science Teaching?

As has been found in previous studies, our results suggest that the introduction of socio-scientific issues was a new approach which teachers found challenging and different to their usual practice, but worthwhile (Tidemand and Nielsen 2017). Overall, our findings show that the teachers in our study were able to use teaching strategies presented at a workshop to create lessons of their own based on new socio-scientific cases they developed, and reflect on the possibilities and challenges of introducing this topic in science teaching. Also, the teachers felt that the introduction of SSI approaches generated enthusiasm and learning in their students.

We found, as is often the case with the introduction to new techniques in teaching, that the ways in which teachers appropriated SSI approaches into their science teaching varied greatly depending on their starting points with regards to their content knowledge but also, particularly, their views on the nature of science learning (Evagorou and Puig Mauriz 2017). In this sense, we propose that the successful and sustained incorporation of SSI approaches depends on two factors; on the one hand, having strong content knowledge and, on the other, valuing learning science in more active and contextualized ways.

Following our model, we would therefore argue that Clara, who did not consider herself as a science content specialist, was able to incorporate the initiatives suggested in the workshop to kindergarten level teaching since her view of teaching more generally (and science teaching as a part of that) already incorporated the idea that students should learn in a contextualized and authentic way. We saw how (with the help of the researchers) she was able to create a whole science unit on the topic of soil and water (that is, the topic of the workshop) that incorporated the SSI approaches and repeat the activity in the following year. However, we believe these changes in her practice were not widened to other science topics over time due to a lack of science content knowledge, for which she would have needed a way to either deepen it, or have access to further 'step by step' teaching materials or ongoing support from the researchers.

Clara's experience was considerably different to Julia who, despite having solid foundational content knowledge in chemistry, did not integrate the SSI approach as a central part of her regular lessons but rather used them to 'sugarcoat' the lessons and make the topics more appealing to her students. One possible explanation for this outcome would be that as her views on the nature of science learning prioritized the direct acquisition of facts, formulas and concepts, introducing SSI approaches which favour a more ambivalent, complex and 'debatable' view of science did not fit into her 'pedagogical creed', as she put it. We feel that for her to have progressed further into the adaptation or appropriation phases of teacher learning she would have required more profound changes in her views on the nature of science learning.

Lastly, we believe Sofia was able to successfully appropriate the fundamental aspects of SSI approaches due to her solid content knowledge and her views on the nature of science learning as a contextualized and authentic endeavor. Our data shows how Sofia used and adapted several activities and sustained them over time, expanding the approach to new units and subjects, even when no longer having researchers to guide or assist her in co-planning new topics.

Our results suggest that for these changes to endure over time and real appropriation to take place teachers need to truly value the SSI approach and find it to be in tune with their pre-existing mindset regarding science learning. This prerequisite is especially important since using SSI approach can take more time than fact-based, transmissive teaching. Given that the lack of time for planning and covering the national standards is frequently cited as a large barrier to the incorporation of classroom innovations, we agree with others who argue that whenever those new approaches are in tension with a teacher's pedagogical creed, there is a lower chance of them taking root.

Finally, our study points toward several suggestions. First, as in prior studies, we found that having teachers participate first-hand in SSI activities (Evagorou et al. 2014), and then taking home guidelines and student materials, encourages teachers to take a first step towards replicating and adapting these types of activities in their own classrooms, particularly for teachers who are new to the approach (Gray and Bryce 2006). All the teachers implemented some of the workshop activities 'by the book' (be it to support science content knowledge or to suggest novel approaches) and valued being given concrete physical or digital resources to just 'implement' in their classrooms. We would therefore consider this one of the fundamental stages of preparing teachers to implement SSI issues whatever their starting points.

However, we also found that in order to create changes in practice that are sustained over time, it is fundamental to tailor the follow-up to teachers' initial starting points, taking into account both their content knowledge and views of science learning. This approach might allow teachers to strengthen their practice in a stepwise way; starting with the implementation of given or co-planned activities, before moving on to more individual adaptations and full adoption of the different techniques and strategies while, at the same time, helping them to deepen their content knowledge or revising their views of learning when necessary. In turn, this has the potential to increase the engagement, interest and perception of relevance of science education for students.

Therefore, to finish, our study shows that with focused, systematic and strategic interventions, the skills involved in the complex problem-solving undertaken by scientists, citizens and government officials in the face of global challenges might, eventually, make their way to the everyday Argentine science classroom.

References

Caga-anan, K. P., & O'Toole, J. M. (2015). Pre-service science teachers' conceptions of the nature of science and its relationship to classroom practice. *Asia Pacific Higher Education Research Journal (APHERJ), 2*(1).

Cofré, H., González-Weil, C., Vergara, C., Santibáñez, D., Ahumada, G., Furman, M., Podestá, M. E., Camacho, J., Gallego, R., & Pérez, R. (2015). Science teacher education in South America: The case of Argentina, Colombia and Chile. *Journal of Science Teacher Education, 26*(1), 45–63.

Díaz-Moreno, N., & Jiménez-Liso, R. (2013). *Las controversias sociocientíficas como contexto en la enseñanza de las ciencias.* Departamento de Educación. Universidad de Almería.

Dwyer, D. C., Ringstaff, C., & Sandholtz, J. H. (1991). Changes in teachers' beliefs and practices in technology-rich classrooms. *Educational Leadership, 48*(8), 45–52.

España Ramos, E., & Prieto Ruz, T. (2010). Problemas socio-científicos y enseñanza-aprendizaje de las ciencias. *Revista de Investigación en la Escuela, 71*, 17–24.

Evagorou, M., & Puig Mauriz, B. (2017). Engaging elementary school pre-service teachers in modeling a socioscientific issue as a way to help them appreciate the social aspects of science. *International Journal of Education in Mathematics, Science and Technology, 5*(2), 113–123.

Evagorou, M., Albe, V., Angelides, P., Couso, D., Chirlesan, G., Evans, R. H., Dillon, J., Garrido, A., Guven, D., Mugaloglu, E., & Nielsen, J. A. (2014). Preparing pre-service science teachers to teach socio-scientific (SSI) argumentation. *Science Teacher Education, 69*(39–48).

Furman, M., Luzuriaga, M., Taylor, I., Anauati, M. V., & Podestá, M. (2018). Abriendo la "caja negra"del aula de ciencias: un estudio sobre la relación entre las prácticas de enseñanza y las capacidades de pensamiento que se promueven en los alumnos [Opening the "black box" of the science classroom: A study on the relationship between teaching practices and thinking skills that are promoted in students]. *Enseñanza de las Ciencias. Revista de investigación y experiencias didácticas, 36*(2), 81–103.

Furman, M. (2018). La educación científica en las aulas de América Latina [Scientific Education in Latin American Classrooms], Red de Indicadores de Ciencia y Tecnología – Iberoamericana e Interamericana. *El Estado de la Ciencia, 2018*.47–72. Retrieved from: http://www.ricyt.org/files/Estado%20de%20la%20Ciencia%202018/E_2018_EDUCACION_CIENTIFICA.pdf

Gray, D. S., & Bryce, T. (2006). Socio-scientific issues in science education: Implications for the professional development of teachers. *Cambridge Journal of Education, 36*(2), 171–192.

Gulamhussein, A. (2013). Teaching the teachers: Effective professional development in an era of high stakes accountability. *Center for Public Education, 1*, 1–47.

Guskey, T. R. (2002). Professional development and teacher change. *Teachers and Teaching, 8*(3), 381–391.
Lee, H., Yoo, J., Choi, K., Kim, S-W., Krajcik, J., Herman, B. C., & Zeidler, D. L. (2013). Socioscientific issues as a vehicle for promoting character and values for global citizens. *International Journal of Science Education 35*(12), 2079–2113.
Joyce, B. R., & Showers, B. (2002). Student achievement through staff development. In Joyce, B & Showers, B. (2002). *Designing training and peer coaching: Our needs for learning*, VA, USA, ASCD.
Labate, H. (2007). Science education: A (pending) chapter in the curriculum transformation in Argentina. *Prospects, 37*(4), 469–488.
Miller, J. D. (2000). The development of civic scientific literacy in the United States. In *Science, technology, and society* (pp. 21–47). Dordrecht: Springer.
Nielsen, J. A., & Evans, R. H. (2015). Possibilities and limitations from a short methods course in socioscientific teaching. Paper presented at NARST, Chicago, IL, United States.
Pitiporntapin, S., & Topcu, M. S. (2016). Teaching based on socioscientific issues in science classrooms: A review study. *KKU International Journal of Humanities and Social Sciences, 6*(1), 119–136.
Pitiporntapin, S., Yutakom, N., & Sadler, T. D. (2016, December). Thai pre-service science teachers' struggles in using Socio-scientific Issues (SSIs) during practicum. In *Asia-Pacific forum on science learning and teaching* (Vol. 17, No. 2, pp. 1–20). The Education University of Hong Kong, Department of Science and Environmental Studies.
Pugh, K. J., Bergstrom, C. M., Heddy, B. C., & Krob, K. E. (2017). Supporting deep engagement: The teaching for transformative experiences in science (TTES) model. *The Journal of Experimental Education, 85*(4), 629–657.
Pujalte, A., Adúriz-Bravo, A., & Porro, S. (2015). Las imágenes deficiencia en profesoras y profesores de Biología: Entre lo que se dice y lo que se hace. *Revista boletín biológica, 3*, 5–10.
Sadler, T. D. (2011). Situating socio-scientific issues in classrooms as a means of achieving goals of science education. In *Socio-scientific issues in the classroom* (pp. 1–9). Dordrecht: Springer.
Sadler, T. D., Barab, S. A., & Scott, B. (2007). What do students gain by engaging in socioscientific inquiry? *Research in Science Education, 37*(4), 371–391.
Tidemand, S., & Nielsen, J. A. (2017). The role of socioscientific issues in biology teaching: From the perspective of teachers. *International Journal of Science Education, 39*(1), 44–61.
Valverde, G., & Näslund-Hadley, E. (2011). *La condición de la educación en matemáticas y ciencias naturales en América Latina y el Caribe*. Inter-American Development Bank.
Wongsri, P., & Nuangchalerm, P. (2010). Learning outcomes between Socioscientific issues-based learning and conventional learning activities. *Online Submission, 6*(2), 240–243.
Yoon, K. S., Duncan, T., Lee, S. W. Y., Scarloss, B., & Shapley, K. (2016). Reviewing the evidence on how teacher professional development affects student achievement. *Issues & Answers Report, 33*, 1–62.
Zeidler, D. L., & Keefer, M. (2003). The role of moral reasoning and the status of socioscientific issues in science education. In *The role of moral reasoning on socioscientific issues and discourse in science education* (pp. 7–38). Dordrecht: Springer.
Zeidler, D. L., & Nichols, B. H. (2009). Socioscientific issues: Theory and practice. *Journal of Elementary Science Education, 21*(2), 49–58.
Zeidler, D. L., Sadler, T. D., Simmons, M. L., & Howes, E. V. (2005). Beyond STS: A research-based framework for socioscientific issues education. *Science Education, 89*(3), 357–377.

Melina Furman holds a Ph.D. in Science Education from the University of Columbia, United States, and a BSc. in Biological Sciences from the University of Buenos Aires. She is a full-time research professor at the School of Education of the University of San Andrés and researcher at CONICET. She investigates science education from kindergarten to university level, with emphasis on teacher training and educational innovation. Melina also coordinated the Natural Sciences team of the National Teacher Training Institute (INFD, Ministry of Education) from 2017 to 2018, and is the founder of "Expedition Science" which runs summer science camps and other programs to promote science vocations. She wrote and conducted the children's television program "The House of Science" and is a member of the organizing team of TEDxRíodelaPlata and co-founder of "The World of Ideas", a course on innovation for professionals from different disciplines. Melina also coordinated the course of Educational Innovators of CIPPEC and led the Natural Sciences team of the "Escuelas del Bicentenario" Project (IIPE-UNESCO and OEI), an improvement program for schools in vulnerable contexts in Argentina. She has published, among others, the books "Educating curious minds: the formation of scientific and technological thought in childhood" (Santillana Foundation), "The adventure of teaching natural sciences" (Aique) and "Science in the Classroom" (Siglo XXI), in addition to various materials for teachers and students.

Inés Taylor has a degree in Biology from the University of Bristol, England. She completed her postgraduate studies in Education at Kings College London, as part of the Teach First program. Between 2015 and 2016, she was part of Juan Carlos Tedesco's team in the Teaching Improvement Program of the National University of San Martín. As a volunteer and then team leader of the NGO Limited Resources Teacher Training she has led workshops and teacher training conferences in Tanzania, Ghana and Belize. She is currently a research assistant of the Science Education Program at the University of San Andrés, where she is also a tutor in undergraduate and postgraduate courses.

Mariana Luzuriaga holds a degree and postgraduate certificate in Education from the University of San Andrés. She is part of the Science Education Program of the University of San Andrés, participating in various research and professional development projects related to the teaching of Natural Sciences at all educational levels. She is a co-author of academic articles and pedagogical resources, and is involved in the design of training courses located within the framework of the National Teacher Training Institute. She also works as an adjunct professor in the subject "Real Educational Challenges" of the BA Education degree Sciences and as a tutor in the Pedagogical Residence and University Teaching postgraduate degree.

Maria Eugenia Podestá holds a Master in Education from the University of Bath, a postgraduate in Education from the Universidad Católica Argentina and a Bachelor in Biochemistry from the Universidad de Buenos Aires, Argentina. She is a coauthor and compiler of several school improvement books such as "About management and teaching" (Buenos Aires, Ediciones Granica, 2004), "Improving management in schools" (Buenos Aires, Ediciones Granica, 2007), "The Adventure of Teaching Natural Sciences" (Buenos Aires, Aique, 2009), "The Role of the Supervisor in School Improvement" (Buenos Aires, Aique, 2009), "The Brain that Learns" (Buenos Aires, Aique, 2013) and "Wellbeing, Emotions and Learning" (Buenos Aires, Aique 2018). She is the director of the Neurosciences and Education collection of Aique Grupo Editorial, director of the Outreach Area of the School of Education of the University of San Andrés, and co-director of the Science Education Research Program.

Chapter 10
Introducing Model-Based Instruction for SSI Teaching in Primary Pre-service Teacher Education

Anna Garrido Espeja and Digna Couso

10.1 Motivation

Becoming a primary school science teacher implies being familiar with a wide range of pedagogical traditions. Our future teachers receive information, analyze examples and as students experience teaching strategies inspired by different science education ideas. These ideas include the STS (Science, Technology and Society) or competence-based learning paradigms, IBSE (Inquiry-Based Science Education) and MBI (Model-Based Inquiry) methodologies, or the environmental education and STE(A)M (Science, Technology, Engineering, Art and Mathematics) frameworks. The SSI (Socio Scientific Issues) approach, despite its scholarly importance in science education research and its presence in innovative curricula for upper educational levels, is not usually introduced at primary school levels, besides some exceptions (Dolan et al. 2009), and not commonly used in primary school teacher education (Díaz-Moreno and Jiménez-liso 2012). In this case, why should we use SSI as a framework in primary school pre-service teachers' education? What are its potential benefits? We would argue that no matter which methodology or tradition is followed, there are two central messages for our pre-service teachers regarding the teaching and learning of science. The first message refers to the need to shift the focus of science lessons from learning the *products* to participating in the *processes* of science (Duschl and Grandy 2008; Passmore and Svoboda 2012). This implies fostering a cognitive, social and discursive activity in the classroom that

A. Garrido Espeja (✉)
CRECIM, Barcelona, Spain

D. Couso
CRECIM, Barcelona, Spain

Department Didàctica de les Matemàtiques i les Ciències Experimentals, Universitat Autònoma de Barcelona, Barcelona, Spain
e-mail: digna.couso@uab.cat

M. Evagorou et al. (eds.), *Science Teacher Education for Responsible Citizenship*, Contemporary Trends and Issues in Science Education 52,
https://doi.org/10.1007/978-3-030-40229-7_10

resembles that of science. This can be done by involving students in the scientific practices of inquiry, modeling and most importantly, the kind of argumentation that characterizes scientific activity and reasoning (Osborne 2014).

These practices, however, are rarely present in today's schools. One of the main reasons is the *'overemphasis by teachers, curricula, and textbooks on what we know at the expense of how we know (it)'* (Osborne 2010, p. 464), which presents scientific knowledge as indisputable and standardized (Jiménez-Aleixandre 2010). Taking this into account, scientific practices should not be conceived as *new* epistemic and procedural content to be learnt *in addition* to the extensive conceptual content usually dealt with in a typical science subject. On the contrary, what is needed is to re-think, reduce and carefully select the *new* conceptual content to be learnt by participating in those practices. This leads to message two, which refers to the importance of focusing on *'developing an understanding of the major explanatory themes of science'* that are useful to all students (Osborne and Dillon 2008), and can be applied to multiple phenomena and contexts (Izquierdo et al. 1999). This idea is in line with the advocacy for a curriculum around a small set of 'big ideas' in science (Harlen 2010), and the more recent efforts to select 'core disciplinary ideas' within the new STEM US curriculum (NRC 2012). In short, the second message refers to the need to learn more deeply instead of widely.

Agreeing with these two messages implies that whatever the approach one holds for science education, there is a need to involve students in the scientific practices of inquiry, modelling and argumentation to help them develop and learn key scientific ideas, understanding that "*the doing and the learning cannot really be separated*" (Michaels et al. 2008, p. 34). The emergence of SSI as a fruitful possibility for science teaching has usually been associated with using motivating topics to learn argumentation skills (Dawson and Venville 2010; Iordanou and Constantinou 2014; Wu and Tsai 2011), to improve critical thinking and decision making (Albe 2008; Evagorou et al. 2012; Zeidler and Nichols 2009), to contribute to scientific literacy (Kolsto 2001; Sadler and Zeidler 2009), or to apply already-learnt scientific content (Dolan et al. 2009; Sadler and Zeidler 2005; Zohar and Nemet 2002). This is done by discussing and deciding personally or socially relevant controversial issues (Zeidler et al. 2005). However, there is little research that focuses on how SSIs can become useful contexts to actually learn scientific content, and the big ideas of science in particular, while participating in the practices of science, especially at primary school level. In order to do so, primary school teachers would need to learn how to conceive, adapt and/or design SSI lessons in a particular way that emphasizes the conceptual component (key ideas of science), in addition to the epistemic and procedural component (scientific practices). In our proposal we present a teacher education course aimed at introducing SSI at primary school level, within the scientific practices approach, for the promotion of students' learning of the big ideas of science. The aim is to show a case study about the evolution of preservice primary school teachers' skills in designing and teaching SSI lessons from this particular framework.

10.2 Theoretical Framework

Despite the different definitions of SSI, most authors agree that SSI are open-ended and socially controversial (i.e., socially alive), that they have a scientific component while including other disciplines such as politics or economics, and that they imply the evaluation of moral or ethical aspects (Albe 2008; Evagorou et al. 2012; Kolsto 2001). Some typical and well-researched examples of SSI topics are GMO, nanotechnology or climate change. SSI topics addressed to younger students, however, are more local and community based, such as the re-introduction of wild animals, beach sand replacement, or invasive species (Dolan et al. 2009; Domènech and Márquez 2012).

According to the literature, students' participation in SSI allows them to understand the importance of science in everyday life and develop the ability to be critical consumers of scientific information (Kolsto 2001). It also improves their conceptual understanding and interest in science (Albe 2008), encourages participation in discussion and debate, provides a framework for understanding the nature of science (NOS), and develops higher-order thinking skills (HOTS), such as critical thinking and argumentation (Evagorou et al. 2012; Zeidler and Nichols 2009).

SSI can also be useful contexts to promote scientific practices in the classroom, with the most obvious example being the practice of argumentation, clearly present in SSIs due to the controversial nature of these topics and the usual demand for students to argue their positions (Nielsen 2012). In addition, SSIs have been shown to promote conceptual learning, although the construction of this scientific knowledge has generally been considered a preliminary step in the teaching and learning sequence (Dolan et al. 2009). Despite evident links between argumentation and the construction of explanations, it is not so common to find the combination of model-based instruction with SSI, an approach that aims to promote the construction of key scientific ideas while participating in scientific practices within SSI contexts. We use this approach in this paper.

The idea behind this approach is that learning science happens through active and genuine participation in the discursive, cognitive and social practices of science (Duschl and Grandy 2008; Erduran and Jiménez-Aleixandre 2007; Osborne 2014). This implies a profound change of what we should do in the science classroom and how to do it. For instance, it implies promoting the activity of elaborating researchable scientific questions rather than answering factual ones, or scaffolding the process of constructing explanations from the available evidence instead of starting by explaining the scientific theory. It also implies the use of argumentation at a very high level.

In this sense, scientific argumentation is a sophisticated way of arguing that requires a very specific and elaborate form of oral communication (Jiménez-Aleixandre et al. 2003), and of text production (Sanmartí 2003), that does not naturally happen without expert didactical guidance, particularly in a controversial SSI context that involves emotions. In order to help students engage in high-level argumentation that goes beyond the mere presentation of opposing positions, we should foster the discussion around the reasons and use of evidence to support these positions (Adúriz-Bravo et al. 2005), shifting from a 'persuasive dialogue' to a critical 'sense-making dialogue'.

We expect that by promoting this type of 'sense-making' argumentation, pupils would be more likely to use their scientific knowledge. Some authors claim that SSI-based instruction fosters content learning of science (Sadler et al. 2016), or that without content learning that produces 'sense-making', students are not able to effectively participate in argumentation and engage in decision-making (Albe 2008; Nielsen 2012). However, there is less consensus regarding what this content should be, and how it should be acquired.

Some authors have signaled the challenge that deep learning of scientific knowledge for quality argumentation poses for SSI-based instruction (Sadler and Fowler 2006). From the view of these authors, learners need to acquire a knowledge base that is significant in terms of depth, breadth, and organization to allow for quality argumentation, instead of disconnected and compartmentalized concepts. In line with the core disciplinary ideas (NGSS Lead States 2013; NRC 2012), or the 'big ideas' in Science (Harlen 2010), we uphold that students should rather learn a few, important, core School Scientific Models (SSM), which are key models that allow them to predict and explain many different phenomena (Izquierdo et al. 1999), using the same 'rules of the game'. As such, these SSMs encompass school versions of the general ideas behind the fundamental theories in each scientific discipline, being analogous (though not equal) to them. They are developed along schooling by getting more sophisticated in terms of language, representations, details or phenomena to be applied to. Some examples of SSM are the particle model of matter, the Newtonian interactions model or the model of living beings. As an example, the particle model of matter can be used initially to explain the difference between solids and liquids, but it can also be used in later stages to understand why particles of pollution (i.e. PM10) can be in suspension in the air, sustained by the particles that actually form the air. This more sophisticated knowledge could be developed and used when dealing with an SSI about ways to reduce air pollution, to understand both the evidence of pollutants and the mechanisms behind the solutions.

The relation we propose between core SSMs and SSIs is twofold: first, that for adequate participation in SSI students require opportunities to develop a deep understanding of science; and second, that SSI can provide the meaningful contextualization that deep learning of scientific content requires. One could argue, though, that an important model of science (SSM) cannot necessarily be related to phenomena present in SSIs, as these are contemporary and controversial in nature, and often refer to frontier knowledge in science. However, SSMs or core ideas refer to knowledge of disciplines that is so fundamental that it is applicable to many phenomena, including those present in SSIs (i.e. the particle model contains the base knowledge for frontier nanotechnology).

The above ideas indicate that teaching SSI by engaging students in genuine scientific practices to make sense of the data available and build explanations based on key scientific ideas, and SSMs in particular, is a desirable approach for the teaching and learning of science. In order to do so, teachers must be well trained, but initial teacher training is recognized as a problem (Osborne and Dillon 2008), especially when it comes to scientific practices. This is because the complexity of these practices poses great challenges to the initial training of new teachers, who do not usually have the expertise, experience and conceptual and epistemological knowledge for this radically different approach (Reiser 2013). Therefore, on the one hand, future teachers will need support with both the scientific practices and the scientific ideas (SSMs) promoted by those practices (NRC 2007). On the other hand, they require support with the Pedagogical Content Knowledge (PCK) (Shulman 1986) to guide pupils' participation in those practices within an SSI context.

The design case below is a proposal to prepare pre-service teachers to teach SSI at primary school level from the scientific practices framework in order to help pupils build important scientific knowledge, in particular SSMs.

10.3 Design Case

Our investigation takes place in the context of the European Commission funded project (Comenius) *PreSEES*[1] (Preparing Science Educators for Everyday Science), in which we aimed at engaging primary-school pre-service teachers (PTs) with SSI and prepare them to teach SSI in a school. In Spain, science education is part of the general training of pre-service teachers but not part of a specific itinerary. Senior PTs have never heard about socio-scientific issues, and even less about teaching SSI. In this sense, their initial ideas on the topic were expected to be rather limited. To overcome this, an intensive professional development program was designed and implemented within a compulsory subject called 'Practicum IV' (12 ECTS) in the last year (4th year) of the elementary PTs' undergraduate degree. We devoted the last 4 sessions of the subject to the SSI training program, from December 2013 to January 2014, in sessions of 1–2 h (6 h of face to face work +2 h of homework). Seventeen student teachers (12 female, 5 male, range 20–25) participated in this compulsory subject.

Three extra volunteering sessions were held in small groups for PTs to design, implement and reflect on their own SSI lesson plans. They were held from January to April 2014, in sessions of 3 h (9 h of face to face work +10 h of tutored design). Seven student teachers decided to voluntarily participate in these extra sessions and only 4 of them (2 male, 2 female) were able to complete them all. The training program was initially collaboratively designed by international experts within the EU project PreSEES and organized in three modules. These modules were modified and adapted to our national context, with the following three learning objectives:

1. Understand the main characteristics of SSI, scientific practices and SSMs.
2. Plan on teaching SSI from the scientific practices perspective.
3. Teaching practice of SSI.

To achieve these three learning objectives the intervention program had five stages organized in increasing degrees of appropriation of the SSI framework: *experience, analyze, design, implement* and *reflect* (see Fig. 10.1). The figure presents

[1]The PreSEES project is funded by the EU (Comenius/Lifelong Learning) with reference 527,602-LLP-1-2012-1-CY-COMENIUS-CMP from 2012 to 2014.

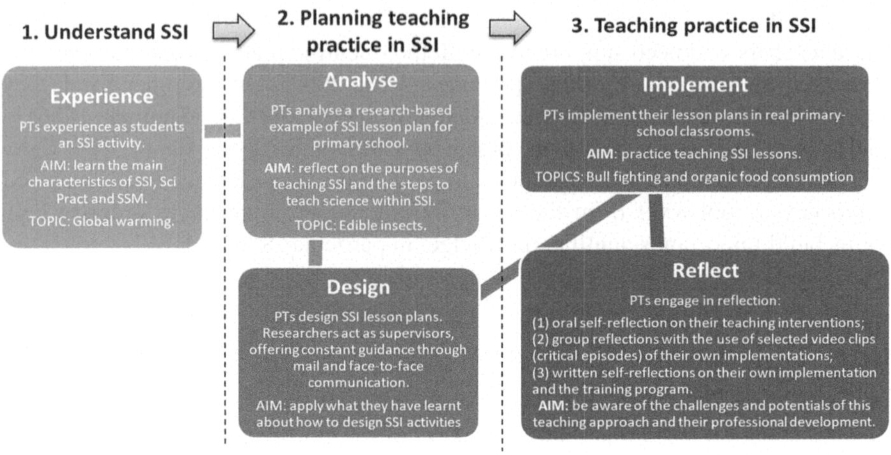

Fig. 10.1 Description of intervention program

the intervention program with three learning objectives (1, 2, 3) with the five stages (experience, analyze, design, implement and reflect). In green coloring, compulsory face to face workshop sessions attended by PTs (n = 17). In blue-purple coloring, volunteer sessions attended. (n = 4–7).

More details of the activities and tasks done in the training program can be found in previous work (Evagorou et al. 2014; Garrido Espeja and Couso Lagarón 2014), and in Fig. 10.2. Figure 10.3 shows a few photographs of different phases of the training program.

10.4 Research

10.4.1 Research Objectives

The focus of this research is on identifying how three pre-service teachers (PTs) that participated in a research-informed initial training on SSIs were able to design and teach SSI lessons within a model-based instruction framework, as well as identifying the potential difficulties that arise in this challenging scenario. The specific research objectives are:

RO1: Portray the evolution of PTs' didactical knowledge[2] and their awareness of their evolution.

[2] We use 'didactic knowledge' here in the sense given by the continental European tradition of science education, which is roughly equivalent to 'pedagogic knowledge' but with an emphasis on the specificities of the science educational field: what and how to teach science education instead of how to teach any subject.

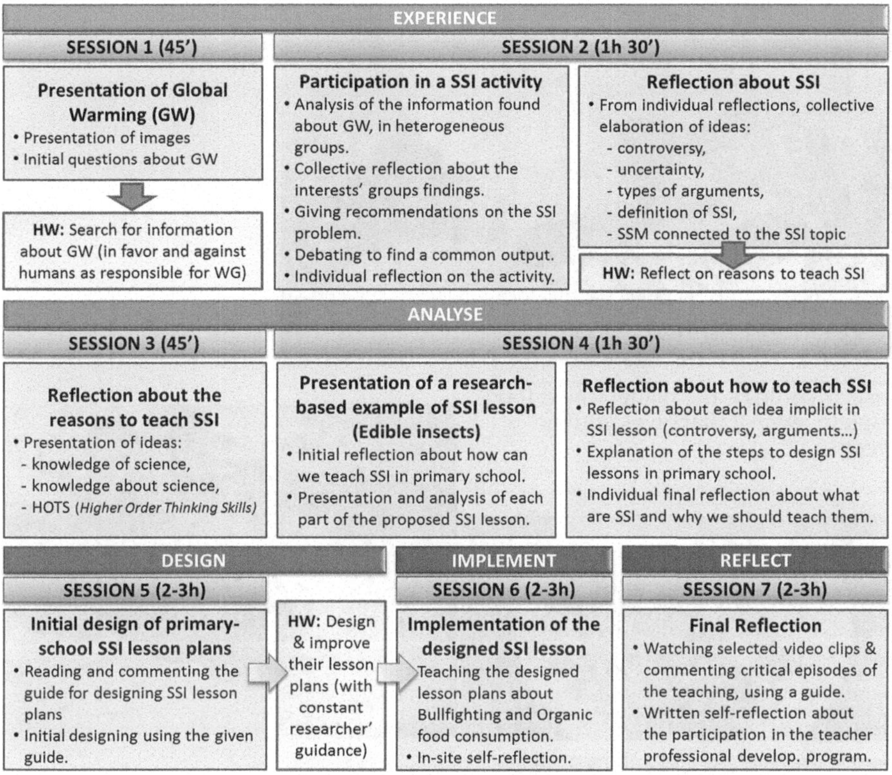

Fig. 10.2 Details of the activities and tasks done in each session of the training program

RO2: Identify PTs' challenges when designing and teaching lesson plans and their awareness of these challenges.

10.4.2 Data Collection and Analysis

From the 17 student teachers that participated in the program, three of them were chosen for the case study: Isabel, Maria and Pol. The reasons were: (1) they showed high interest and good attitudes towards SSI, (2) they had a good level of understanding of SSI according to previous analysis (Garrido and Couso 2015), and (3) they participated in the design, implementation and reflection of their own lesson plans in real elementary schools. Table 10.1 summarizes the profile of the three pre-service teachers used for the case study.

The data collected during the different phases of the intervention includes video recordings of the sessions, oral and written reflections, and different versions of the lesson plans designed by the three selected PTs.

Fig. 10.3 Example of the different phases from the professional development program

Table 10.1 Summary of the three pre-service teachers' profiles chosen for the case study

	Name	Age	Gender	Lesson plan topic	Level	Previous experience
1	Isabel	22	Female	Organic food consumption	3rd grade	Little previous experience on teaching science (only one practicum)
2	Maria	25	Female	Banning of bull-running festival	6th grade	
3	Pol	32	Male			No previous experience in teaching SSI

Table 10.2 Dimensions used to analyze the designed lesson plans of PTs and portray the evolution of their didactical knowledge

Dimension		Description of the dimension
1. Problematizing the topic		A type of overarching question used to introduce and guide the lesson. It should problematize the topic, be interesting/attractive and imply a cognitive challenge for students.
2. Scaffolding		Questions, guidelines or instructions that PTs plan to give to their pupils in order to help them deal with the information found about the SSI topic. PTs should help them face, organize and interpret the different sources, opinions, levels of uncertainty, types of arguments and scientific information gathered.
3.Scientific content	**a. Aim of lesson**	The main objective of the lesson. The learning of the scientific content should be recognized as a key aim of the lesson.
	b. Type	The type of scientific content included in their lesson. PTs should recognize and focus on the key/big scientific ideas (those that help pupils build the main SSMs at the appropriate level) instead of on anecdotic or local ideas.
	c. Development	Tasks designed to help students learn the important scientific content related to the controversial issue. They should include specific tasks to help pupils construct the key scientific ideas and connect them to the SSI topic.
4. Assessment		The type of assessment planned. PTs should include in their lesson plans specific and formative ways to assess students' learning of the SSI topic and the SSM, and their engagement in scientific practices such argumentation.

To portray how ideas about teaching science and other didactical aspects changed during the scaffolded design process (RO1), we compared PTs' initial and final teaching designs, and analyzed their oral and written reflections about their learning process. This aimed to identify different didactical aspects that changed (or did not change) in their designs, and the ideas in their reflections that were coherent with the didactical aspects already identified in their lesson plans. The four dimensions analyzed to determine evolution of the PTs' didactical knowledge are described in Table 10.2.

To identify the difficulties pre-service teachers faced when trying to improve their lesson plans and teaching, and to determine whether they were aware of these difficulties (RO2), we looked for critical episodes in their reflections (oral and written) to construct an interpreted narrative.

10.4.3 Results

In our case-study we identified results on three main dimensions: the evolution of PTs' SSI lessons plans and PTs' awareness of their own progress (RO1); the inclusion of scientific content in SSI lessons (RO2); and the dialogic role of PTs in SSI discursive activities (RO2).

10.4.3.1 Evolution of PTs' SSI Lessons Plans and PTs' Awareness (RO1)

Pre-service teachers' designs changed during the designing process regarding the four didactical aspects we analyzed. This was done by: (1) including a real problematization of the lesson, (2) scaffolding students' work adequately with diverse information, (3) including the scientific content (as an aim of the lesson, recognizing the important key scientific ideas or including specific tasks to construct scientific knowledge), and (4) using specific and adequate ways to assess pupils' learning. Extracts of initial and final designs of PTs' lesson plans are presented to provide evidence of changes in each of the dimensions (Table 10.3).

Pre-service teachers were very aware of these changes, and they actually reflected on their own professional learning during the training intervention, recognizing the usefulness of the participatory lessons attended and highlighting their increased capability to design more innovative science lessons:

> '*I have learnt to be a teacher.*' (Pol).
> '*I have learnt to do science from another perspective, in a different way* [different from the traditional model they used to follow].' (Maria).

They also expressed that they feel capable of teaching SSI in primary school, highlighting the real possibility of relating SSI and the scientific content:

> '*I have seen that it is possible to teach SSI in primary school.*' (Pol).
> '*Now I am able to design and implement SSI activities in primary schools by myself.*' (Maria).
> '*I have learnt that doing SSI is not an extra activity from all the topics we have to teach. We have to relate the SSI with the scientific curriculum.*' (Isabel).

Nevertheless, they recognized some important aspects, such as the **need to master the scientific content** and **know about the SSI topic** in order to successfully teach SSI:

> '*I have learnt that the teacher should become an expert on the topics [scientific and controversial] in order to successfully communicate knowledge.*' (Maria).

These PTs were not used to designing lesson plans, and when planning they did not consider possible solutions for the planned problems nor include enough guidance in the designed tasks. It seems that they expected to solve the majority of the problems and difficulties at the time of the practice, giving in-situ responses, which made the implementations more complex and demanding than necessary. By the end of the program, they recognized their new ability to predict possible problems in the classroom:

> '*I have learnt [...] to know the possible problems that can arise with elementary school students.*' (Pol).

Similarly, PTs were not aware of the need of unveiling the information to students at the beginning of the training. They tended to expect pupils to face the original information (unmodified), such as news or websites, with a very high level of content and language for students. They also did not plan how to guide and scaffold students' work with information to help them make sense of it. After receiving the

Table 10.3 Summary of key changes in the designed lesson plans

		Initial designs	Final designs
1.Problematizing the topic		No overarching question used, just a title on the topic: *"The use of animals in traditional festivities"*	Overarching question that problematizes the topic is used: *"Should animals be used in traditional festivals? Should we continue doing the festival or should it be banned?"*
2. Scaffolding		Vague explanation about how they will help pupils deal with information: "*I will ask students about what they had read*"	Specific tasks with adequate scaffolding to help pupils deal with the information found: "*In heterogeneous groups, each student will have to complete a document with the following questions: Are they in favor or against? What type of arguments are they using?*"
3. Scientific content	**a. Aim of lesson**	Scientific content is not recognized as an aim of the lesson. Other aims mentioned (i.e. *identify different points of view*).	Scientific content is specifically included as an aim of the lesson: *"Learn about the role of stimulus-response: how the nervous system works, the 5 senses and the kind of response given in the form of pain, stress, heartbeat, etc."*
	b. Type	The scientific content included is anecdotic and does not build on any SSMs: "*Students can learn that when bulls are taken out of their natural environment they feel bad.*"	Awareness on the importance of focusing on the big scientific ideas and inclusion of adequate SSMs that connect with the SSI topic: "*We don't want students to become experts in bulls, but to know that this idea [stimulus-response] is general for all living beings.*"
	c. Develop-ment	No inclusion of tasks to teach the scientific content and recognition of the challenge to do so: "*We focus so much on the different opinions, but the scientific content remains unlinked and I don't know how to do it.*"	Inclusion of adequate tasks to help students learn SSMs: "*Students will experiment with eyes closed in pairs and then draw where the information travels through the body. Then, we will discuss the drawings in a big group and agree on the best drawing (explaining how the nervous system and the relation function work).*"
4. Assessment		No assessment is included in the designed lessons.	An adequate assessment task is proposed to evaluate pupils' knowledge of the SSI topic, scientific content and argumentation skills: "*As you have seen, there are people who believe that animals do not suffer because they are not human beings but there are others who believe the opposite. Taking that into account, what do you think about animal testing for making cosmetics? Explain your opinion to someone who doesn't know about it, taking into account what you know about the nervous system and the relation function*".

support of the researchers, PTs developed some strategies to help their students and by the end of the training they recognized how important it is to help students deal with the complex and abundant information that can be found around SSI.

10.4.3.2 The Inclusion of the Scientific Content in SSI Lessons (RO2)

In addition to the general problems of the teaching practice, the SSI context also poses particular challenges to novice teachers. Finding ways to link the scientific content with the SSI topic was one of the most important concerns and a big challenge for all PTs, during both the designing process and during implementation of the teaching lesson. They expressed their worries on multiple occasions regarding the challenge of including the scientific content in the lesson and meaningfully connecting it to the SSI topic:

> '*I think it is very interesting that we work on the scientific content but I don't know how to do it. We focus so much on the different opinions and arguments but this part [scientific content] remains unlinked.*' (Maria, design session).

They also found it challenging to make students understand that it is necessary to be well informed and base their opinions on the scientific knowledge available to have well-founded opinions and adequate positions regarding the topic. Helping students use the scientific evidence when justifying their opinions was one of the greatest challenges these PTs faced:

> '*The most difficult part for me has been to relate the scientific content with controversy [SSI], and make students see that behind that controversial issue there is some scientific content that needs to be understood in order to justify their opinion. We had to ask them questions and make the connection more explicit, and despite doing so, they didn't do it.*' (Maria, reflection session).
>
> '*In the end most students positioned themselves against the bull festival, but their reasons were not based on scientific evidence.*' (Pol, in-site self-reflection).

Regardless of these important challenges, PTs were able to critically reflect on their implementation and they realized of the achievements involved, such as being able to include key scientific ideas in the lesson connected to the SSI:

> '*What I have learnt about SSI is to connect scientific topics with SSI.*' (Isabel, written self-reflection).

They even thought of possible new ways to help students use their scientific knowledge in an SSI lesson, for example, by asking them to use scientific evidence to justify their ideas and opinions, or initiating the lesson with a question that requires scientific knowledge to be solved instead of a statement to be learnt:

> '*When that girl was saying that the bull was suffering and the other boy replied, maybe we should have asked: 'how do you know if the bull feels anything?*'' (Pol, reflection session).
>
> '*Before we did the activity of the stimulus-response system, maybe we should have asked them from the beginning: 'do they suffer or not? Why?*'' (Maria, reflection session).

10.4.3.3 The Dialogic Role of Pre-service Teachers in SSI Discursive Activities (RO2)

In the literature in science education the interactions between teachers and students have been largely studied, in particular regarding the dialogic role of these interactions (Mortimer and Scott 2003). In our study we have seen that pre-service teachers also experienced great difficulties in knowing what type of dialogic role they should play when teaching SSI. On the one hand, they were afraid of convincing students to think one way by unconsciously suggesting their own position on the topic. Therefore, they didn't give enough information about SSI or the necessary instructions to students. On the other hand, they had an excessively active role in the dialogic activity (i.e. the debate), strongly guiding the discussion and making unnecessary explanations because they were afraid of having to deal with uncontrolled situations or bad-manners in discussions.

> '*At the beginning I tried not to influence their opinions and let them speak. That's why I didn't communicate the questions well enough.*' (Maria, reflection session).
>
> '*I was controlling the debate too much, but I wanted everybody to speak... Now I see, I should give them more freedom but I am not used to it.*' (Isabel, reflection session).

They also used a traditional transmissive teaching methodology when introducing scientific content (offering information, facts and answers), instead of guiding students in using scientific knowledge in their argumentation or helping them reformulate ideas on their own:

> '*I was surprised at their good ideas but I was afraid of not having enough time, that's why I gave them the answers.*' (Isabel, in-site self-reflection).

Nevertheless, PTs were able to critically reflect on their dialogic role, recognizing important aspects to take into account to better guide the debate:

> '*I have learnt the importance of good questions. I have learnt to listen more to students and to value any intervention.*' (Maria, written self-reflection).

10.5 Summary and Conclusions

The teacher education programme proposed here has been successful in terms of PT's professional development, as shown by the improvement and innovation quality of their designed lesson plans, the critical reflection on their implementations and the awareness of their own learning, and their recognition of the positive impact the training program had on their professional development. As shown, final versions of the lessons improved in four important aspects: (1) the use of an overarching question that problematizes the lesson; (2) the inclusion of the scientific content in the aims of the lesson, both in the form of key SSMs to be built through specific modelling tasks; (3) the presence of adequate scaffolding that supports pupils' mastering of the information found; and (4) the planning of adequate formative assessment. Some particular challenges arose during the process, in particular regarding: (1) the need to anticipate the problems that could arise during the implementation; (2) the type of support necessary for primary-school students to deal with the large amount and complexity of information; (3) the connection between the SSM and the SSI topic to promote learning of key ideas while learning SSI; and (4) the dialogic role of the teacher.

Regarding the need to anticipate possible problems during the design phase, we have observed that despite being complex and demanding for PTs, asking them to design innovative learning sequences with the SSI approach has proved to be a very powerful teacher training strategy not sufficiently used in teacher education programs. As has been shown in this case study, PTs were aware of the importance of planning possible solutions to the expected problems in advance. The challenge posed by the SSI approach helped PTs realize that anticipation is better than their tendency to solve problems 'on the way'.

Properly dealing with information is one of the main competencies that SSIs seek to promote, but finding ways to support pupils with this process in elementary classrooms is a particular challenge. Pupils needed help to select, understand and/or use various sources of information. PTs were not aware of this challenge at the beginning of their training, despite having undergone an SSI activity as students themselves and having reflected on different examples of SSI lessons. This shows that teacher education in the context of SSI has to give importance to helping PTs develop strategies and tools to help students understand and organize information in adequate ways.

Of all the difficulties, it turned out to be particularly challenging for PTs to include scientific content in a non-simplistic or superficial way. Making students understand the importance of mastering the scientific knowledge, and promoting the use of scientific knowledge in their argumentation to form well-founded opinions was hard for PTs, which is in line with what has been seen in other studies on SSI (Nielsen 2012). PTs encountered significant challenges when looking for ways to effectively promote evolution of the initial ideas of students instead of just transmitting the scientific content as fully-formed knowledge. This was probably due to PTs own lack of both sufficient pedagogical content and subject matter knowledge. They also had difficulties designing activities that seek pupils' construction of the main key ideas of the SSM instead of just teaching vocabulary or details. However, they were able to identify their own limitations and think of ways to overcome these problems in future situations, showing mature and critical reflection (Berland and Reiser 2011). The very detailed guidance and constant scaffolding from researchers helped PTs become aware of the need to link the learning of scientific content with the learning of other important abilities associated with SSI activities, such as HOTS. They developed a vision of SSI teaching as a fruitful context for teaching science in an innovative way, rather than as an additional methodology to apply already known scientific content.

The dialogic role of teachers in an SSI lesson, particularly in discursive activities such as the final debate, was also a problematic aspect for PTs. There was a contradiction between their ideas about how to teach and their actual practice, and PTs ended up 'guiding' more than planned in some situations, or not giving enough guidance to promote genuine and active participation in scientific practices. For instance, teachers were very traditional and teacher-centered when teaching scientific content and played too active a role in the debates, constantly giving speaking turns and making explanations to clarify students' interventions. However, they did not guide enough when engaging students in argumentation or modeling, such as by helping them use the scientific knowledge in their arguments or reformulate their ideas in their own words.

Despite all these challenges, future teachers were able to critically reflect on their own progress. In this regard, they recognized that the professional development program on SSI was particularly fruitful, and expressed their desire that all teachers, both pre-service and in-service, could receive training on SSI in order to improve their pedagogical model. In our view, reasons behind these positive results are related to different factors: the intrinsic potential that SSI have regarding teacher education, the particular characteristics of the teacher education program we followed, but also the particular SSI approach used that includes model-centered instruction. In the analysis of this experience we have identified aspects that should be taken into account when designing and implementing pre-service teacher education programs. These aspects pose implications for teaching and research, which are discussed in the following section.

10.6 Implications for Teaching and Research

From our analysis we have seen that learning to be a teacher from an SSI approach is a challenge for PTs, but it promotes more innovative teaching practices than other possible approaches used in pre-service teacher education, such as school-focused approaches aimed at 'doing the lesson' which inhibit student argumentation (Jiménez-Aleixandre et al. 2000). We consider that there are some intrinsic characteristics of SSI contexts that promote these innovative teaching practices. For instance, the fact that SSI topics are current controversies emphasizes the need to introduce the content in a problem-based manner rather than as a fact. In addition, SSI emphasizes positioning and making decisions in an evidence-based manner, which requires students to deal with diverse information in diverse formats (news, reports, etc.) that may not be easily understood. This implies that students are more likely to question the scientific knowledge, connect it with daily and contextualized situations and use it to solve real problems, which makes it harder for teachers to keep a traditional and teacher-centered approach.

Nevertheless, learning to teach SSI is not a simple task that can be easily implemented within any teacher education program. From our research, we have identified two important characteristics. First, the importance of letting PTs experience SSI lessons first-hand (as students), in order for them to have the experience of facing socially alive situations, dealing with complex information, learning to build good arguments in an evidence-based manner, and positioning themselves in a controversial, open-ended topic. This first-step of experiencing is crucial because future teachers usually lack alternative learning experiences to traditional or transmissive ones, which makes it difficult for them to imagine other types of didactic approaches and how to bring them to the classroom (Avraamidou and Zembal-Saul 2010; Martínez Chico et al. 2014). However, we are aware that making PTs participate in adequate and innovative teaching methodologies is not enough to learn all the complexities of teaching science and SSIs in particular. Thus, the second characteristic is the importance to follow a complete cycle of Design-Implementation-Reflection (D-I-R) in the context of a subject that allows teaching practice. This means that PTs should have to face the challenge of designing real lesson plans, bringing them to real elementary-school classrooms with real students and critically reflecting on the whole process. They should reflect on positive and negative aspects, their own challenges, the ways they overcame them, the limits of their teaching practice, possible improvements for next time and their own evolution during the process. This intense DIR cycle can succeed if it is adequately scaffolded, that is, both guided and supervised by teacher educators throughout the entire process in a trusting environment. The literature has innumerable accounts of the potential of this tutor-mentee structure. We just want to add that providing continuous and personalized support, and focusing the reflection-on-practice by video-based discussions is promising, but requires high investments in time and resources.

To end with, we would like to reflect on the particular type of teaching and learning approach given to SSI in this teacher education course. The literature has largely advocated for the SSI approach as a framework for the development of transversal competences and HOTS such as active citizenship, critical thinking and argumentation, and learning to apply scientific knowledge (Evagorou et al. 2012; Nielsen 2012; Zeidler and Nichols 2009). In these proposals, the learning of the scientific content is done before (as a prerequisite for) SSI involvement, usually via IBSE. However, we advocate for a view of SSI as a fruitful scenario for the construction and re-construction of scientific knowledge. This implies that SSI can become a very fruitful teaching and learning approach within the scientific practices framework for the learning of core ideas in science (NRC 2012). Our standpoint is that the SSI context and argumentation activity gives purpose to the learning of the scientific model (Andriessen 2006; Ryu and Sandoval 2012), and that knowledge produced by school science cannot (and should not) be separated from the social practices of its production (Adúriz-Bravo 2008; Ryu and Sandoval 2012).

In our proposal, we embed model-centered instruction with SSI for this purpose by introducing SSI to build key School Scientific Models (SSM). In doing so, PTs professional development becomes more complete, in the sense that they learn to master the content knowledge in addition to internalizing a way of teaching that promotes both construction of knowledge and evidence-based argumentation via dialogic teaching. With this approach, pre-service teachers learn to teach in a way that does not focus on 'doing the lesson', as they are used to, but on 'sense-making' of ideas in the classroom that let students develop quality argumentation and give coherent answers to current and controversial topics. From this teaching approach, the selection of SSIs topics should take into account not only their motivational and engaging nature, but also their connection with the core SSMs that all students should master. In this study we have seen that a proper selection of this content, in relation to the selection of the SSI topic, is crucial: it is one of the biggest challenges when designing SSI lessons from the perspective of scientific practices.

References

Adúriz-Bravo, A. (2008). Un modelo de ciencia para el análisis epistemológico de la didáctica de las ciencias naturales. *Perspectivas Educativas, 1*, 13–39.

Adúriz-Bravo, A., Bonan, L., Galli, L. G., Chion, A. R., & Meinardi, E. (2005). Scientific argumentation in pre-service biology teacher education. *Eurasia Journal of Mathematics, Science and Technology Education, 1*(1), 76–83.

Albe, V. (2008). Students' positions and considerations of scientific evidence about a controversial socioscientific issue. *Science & Education, 17*(8–9), 805–827.

Andriessen, J. (2006). Arguing to learn. In R. K. Sawyer (Ed.), *The Cambridge handbook of the learning sciences* (pp. 443–459). Cambridge: Cambridge University Press.

Avraamidou, L., & Zembal-Saul, C. (2010). In search of well-started beginning science teachers: Insights from two first-year elementary teachers. *Journal of Research in Science Teaching, 47*(6), 661–686.

Berland, L. K., & Reiser, B. J. (2011). Classroom communities' adaptations of the practice of scientific argumentation. *Science Education, 95*, 191–216.

Dawson, V. M., & Venville, G. (2010). Teaching strategies for developing students' argumentation skills about Socioscientific issues in high school genetics. *Research in Science Education, 40*(2), 133–148.

Díaz-Moreno, N., & Jiménez-liso, M. R. (2012). Las controversias sociocientíficas : temáticas e importancia para la educación científica. *Eureka, 9*(1), 54–70.

Dolan, T. J., Nichols, B. H., & Zeidler, D. L. (2009). Using socioscientific issues in primary classrooms. *Journal of Elementary Science Education, 21*(3), 1–12.

Domènech, A. M., & Márquez, C. (2012). Students' opinions about a SSI: Perspectives refered in their arguments about bears' reintroduction. In C. Bruguière, A. Tiberghien, & P. Clément (Eds.), *E-book proceedings of the ESERA 2011 conference* (pp. 31–38). France: Lyon.

Duschl, R. A., & Grandy, R. E. (2008). *Teaching scientific inquiry: Recommendations for research and implementation*. Rotterdam: Sense Publishers.

Erduran, S., & Jiménez-Aleixandre, M. P. (2007). *Argumentation in science education. Perspectives from classroom-based research*. New York: Springer.

Evagorou, M., Jiménez-aleixandre, M. P., & Osborne, J. (2012). "Should we kill the Grey squirrels?" a study exploring students' justifications and decision-making. *International Journal of Science Education, 34*(3), 401–428.

Evagorou, M., Albe, V., Angelides, P., Couso, D., Chirlesan, G., Evans, R., et al. (2014). Preparing pre-service science teachers to teach socio-scientific (SSI) argumentation. *Science Teacher Education, 69*, 39–47.

Garrido, A., & Couso, D. (2015). Socio-scientific issues (SSI) in initial training of primary school teachers: Pre-service teachers' conceptualization of SSI and appreciation of the value of teaching SSI. *Procedia – Social and Behavioral Sciences, 196*, 80–88.

Garrido Espeja, A., & Couso Lagarón, D. (2014). Les controvèrsies socio-científiques (SSI) en la formació inicial de mestres de primària: Anàlisi de l'aprenentatge, l'auto-eficàcia i l'aplicació real a l'aula. In *Special issue for the VIII international conference on university teaching and innovation CIDUI 2014* (Vol. 2. pp. 1–10).

Harlen, W. (2010). *Principles and big ideas of science education*. Hants: Gosport.

Iordanou, K., & Constantinou, C. P. (2014). Developing pre-service teachers ' evidence-based argumentation skills on socio-scientific issues. *Learning and Instruction, 34*, 42–57.
Izquierdo, M., Espinet, M., García, M. P., Pujol, R. M., & Sanmartí, N. (1999). Caracterización y fundamentación de la ciencia escolar. *Enseñanza de Las Ciencias*.
Jiménez-Aleixandre, M. P. (2010). *10 ideas clave. Competencias en argumentación y uso de pruebas*. Barcelona: GRAO.
Jiménez-Aleixandre, M. P., Bugallo Rodríguez, A., & Duschl, R. A. (2000). "Doing the lesson" or "doing science": Argument in high school genetics. *Science Education, 84*(6), 757–792.
Jiménez-Aleixandre, M. P., Caamaño, A., Oñorbe, A., Pedrinaci, E., & de Pro, A. (2003). Enseñar ciencias. Barcelona: GRAO.
Kolsto, S. D. (2001). Scientific literacy for citizenship: Tools for dealing with the science dimension of controversial socioscientific issues. *Science Education, 85*(1), 291–310.
Martínez Chico, M., López-Gay Lucio-Villegas, R., & Jiménez Liso, M. R. (2014). ¿Es posible diseñar un programa formativo para enseñar ciencias por Indagación basada en Modelos en la formación inicial de maestros? Fundamentos, exigencias y aplicación. *Didáctica de Las Ciencias Experimentales Y Sociales, 4379*(28), 153–173.
Michaels, S., Shouse, A. W., & Schweingruber, H. A. (2008). *Ready, set, science!* Washington, DC: The National Academies Press.
Mortimer, E. F., & Scott, P. H. (2003). *Meaning making in secondary science classrooms* (McGraw-Hill Education, Ed.). Maidenhead: Open University Press.
NGSS Lead States. (2013). Next generation science standards: For States, by States. Washington, DC: National Academies Press
Nielsen, J. A. (2012). Science in discussions: An analysis of the use of science content in socioscientific discussions. *Science Education, 96*(3), 428–456.
NRC. (2007). *Taking science to school: Learning and teaching science in grades K-8*. Board on Science Education, Center for Education. Division of Behavioral and Social Sciences and Education. Washington, DC: The National Academies Press.
NRC. (2012). *A framework for K-12 science education. Practices, crosscutting concepts and core ideas*. Washington, DC: The National Academies Press.
Osborne, J. (2010). Arguing to learn in science: The role of collaborative, critical discourse. *Science, 328*, 463–466.
Osborne, J. (2014). Teaching scientific practices: Meeting the challenge of change. *Journal of Science Teacher Education, 25*(2), 177–196.
Osborne, J., & Dillon, J. (2008). *Science education in Europe: Critical reflections*. London: Nuffield Foundation.
Passmore, C. M., & Svoboda, J. (2012). Exploring opportunities for argumentation in modelling classrooms. *International Journal of Science Education, 34*(10), 1535–1554.
Reiser, B. J. (2013). What professional development strategies are needed for successful implementation of the next generation science standards? *Invitational Research Symposium on Science Assessment*.
Ryu, S., & Sandoval, W. A. (2012). Improvements to elementary children's epistemic understanding from sustained argumentation. *Science Education, 96*(3), 488–526.
Sadler, T. D., & Fowler, S. R. (2006). A threshold model of content knowledge transfer for socioscientific argumentation. *Science Education, 90*, 986–1004. https://doi.org/10.1002/sce.20165.
Sadler, T. D., & Zeidler, D. L. (2005). The significance of content knowledge for informal reasoning regarding socioscientific issues: Applying genetics knowledge to genetic engineering issues. *Science Education, 89*(1), 71–93.
Sadler, T. D., & Zeidler, D. L. (2009). Scientific literacy, PISA, and socioscientific discourse: Assessment for progressive aims of science education. *Journal of Research in Science Teaching, 46*(8), 909–921.
Sadler, T. D., Romine, W. L., & Topçu, M. S. (2016). Learning science content through socioscientific issues-based instruction: A multi-level assessment study. *International Journal of Science Education, 38*(10), 1622–1635.

Sanmartí, N. (2003). *Aprendre ciències tot aprenent a escriure ciencia* (Edicions 62, Ed.). Barcelona.

Shulman, L. S. (1986). Paradigms and research programs in the study of teaching: A contemporary perspective. In M. C. Wittrock (Ed.), *Third handbook of research on teaching*. New York: Macmillan.

Wu, Y. T., & Tsai, C. C. (2011). High school students' informal reasoning regarding a socio-scientific issue, with relation to scientific epistemological beliefs and cognitive structures. *International Journal of Science Education, 33*(3), 371–400.

Zeidler, D. L., & Nichols, B. H. (2009). Socioscientific issues: Theory and practice. *Journal of Elementary Science Education, 21*(2), 49–58.

Zeidler, D. L., Sadler, T. D., Simmons, M. L., & Howes, E. V. (2005). Beyond STS: A research-based framework for socioscientific issues education. *Science Education, 89*(3), 357–377.

Zohar, A., & Nemet, F. (2002). Fostering students' knowledge and argumentation skills through dilemmas in human genetics. *Journal of Research in Science Teaching, 39*(1), 35–62.

Dr. Anna Garrido Espeja is a researcher at the Research Centre CRECIM at Universitat Autònoma de Barcelona and she has worked as a teacher educator at Universitat de Barcelona and Universitat de Girona, Spain. She has participated in several national and European projects regarding Socio-Scientific Issues in science teaching, pre-service teacher education and models and modelling in science education. In her PhD work, she focused on the learning of models and the participation of pre-service teachers in the scientific practice of modelling.

Dr. Digna Couso is a reader in science education and the director of the Research Centre for Science and Mathematics Education (CRECIM) at Universitat Autònoma de Barcelona, Spain. Her research interests are models and modelling for the teaching and learning of science, participatory design-based approaches to teacher education and equity and gender issues in STEM Education, with a special interest in ICTs (Information and communication technologies). She has lead and participated in a wide range of national and European projects and is a referee of some prestigious journals in the field.

Chapter 11
Re-thinking the Integration of Socioscientific Issues in Life Sciences Classrooms Within the Context of Decolonising the Curriculum

Ronicka Mudaly

11.1 Background

During recent decades, the global science education landscape has reverberated with the call to integrate socioscientific issues (SSIs) into science education as a way towards responsible citizenship. This was embedded in a shift towards a humanistic science education, which witnessed a departure from a technicist towards a dynamic, democratic, socially responsive science education. This was marked by a disruption of strong (Bernsteinian) curriculum boundaries which insulate "pure" science as a separate entity, which occupies a superior position in the knowledge repository. The notion of being a value-free scientist who is ensconced in a university laboratory, and whose discourse is permeated purely with reductionist processes, is being criticised. The need for a science education which troubles social issues and generates uncertainty and discomfort, in a quest to locate opportunities for substantive, helpful social change, has paved the way for interdisciplinary studies. Increasingly, a transformative epistemology has underpinned ways in which science education and society intersect. Within the South African context, complicated conversations (Pinar 2004) about transformative epistemology in general, and curriculum transformation in particular, has dominated debates in public and private spheres. The exclusion of knowledge systems of indigenous people from mainstream education, and the role of education in reproducing knowledge hierarchies which are dominated by Euro-Western frameworks, permeated arguments for Africanising education. The principles of the National Curriculum Statement, which underpin the school curriculum, are imbued with civil, economic, cultural and social human rights, which are enshrined in the Constitution. However, although South Africa is in its third decade of a post-apartheid political order, Euro-Western

R. Mudaly (✉)
School of Education, University of KwaZulu – Natal, Durban, South Africa
e-mail: mudalyr@ukzn.ac.za

M. Evagorou et al. (eds.), *Science Teacher Education for Responsible Citizenship*, Contemporary Trends and Issues in Science Education 52,
https://doi.org/10.1007/978-3-030-40229-7_11

epistemic traditions persist as a resilient feature of school and post-school curricula. The lack of transformation in Basic and Higher Education Institutions is evident due to their continued marginalisation of non-Western ideological frameworks (Heleta 2016; le Grange 2016).

Researchers such as le Grange (2018) emphasise that colonial and apartheid margins of division continue to influence curricula. University and school science curricula have been complicit in promoting Euro-Western worldviews as universal. In the face of this social reality, the constitutional pillars of social justice, human rights and equity become ideological rhetoric. Within this context, South African society has witnessed unprecedented protests by students, who have become increasingly disconnected from, and discontented with education, and this has precipitated calls for decolonising the curriculum. The broad social challenge, then, is the lack of articulation between the principles of social justice contained the National Curriculum Statement, and the implemented curriculum, the latter which drills students "with en vogue epistemologies" (Magill and Rodriguez 2014, p. 209) and reproduces the unequal social order.

In this study, ways in which teachers embed SSIs into a unit of work in the curriculum, towards a more just and socially relevant epistemology, are explored. First, a rationale for SSI-based instruction is offered. Next, the argument for drawing on theoretical constructs from critical pedagogy is presented. This is followed by a description of the qualitative methodological approach, the results and analysis thereof, and finally, the concluding remarks.

11.2 Rationale for Integrating SSI into the Curriculum

The interdependence between (science) education and society is summed up succinctly by Magill and Rodriguez (2014, p.210) who cite Dewey and Durkheim's perspective of education as including "classroom experience, reflective thinking, further interaction with the curriculum, and rethinking based on what is encountered…" They conclude by asserting that "a student's curriculum is their life and life is a student's curriculum." Giroux (2011, p.3) adds: "Education is fundamental to democracy and no democratic society can survive without a formative culture shaped by pedagogical practices capable of creating the conditions for productive citizens who are critical, self-reflective, knowledgeable, and willing to make moral judgements and act in a socially responsible way". The marginalisation of alternative knowledge systems, the national call to decolonise the curriculum, and ideas about the role of democratic education within society are interconnected, and provide the rationale for SSI-based instruction as a response.

Integrating socioscientific issues into the science curriculum has the potential to connect science to humanity (Talens 2016). Engaging learners in science education which is interrelated with social issues increases the personal meaningfulness and relevance of science to learners (Mnguni 2017; Mudaly 2011; Onwu and Kyle 2011; Talens 2016; Zeidler and Nichols 2009). However, relatively few teachers incorporate SSI in their curricula (Klosterman et al. 2012). Many teachers retreat to the default pedagogy of transmission of science content because of the pressure of completing the syllabus (ibid, 2012). Others find teaching SSI controversial and exhibit negative attitudes to it (Osbourne et al. 2002).

In South Africa, professional development of teachers is linked to school curriculum documents. Specific aims are stipulated in the Curriculum Assessment and Policy Statement (CAPS) for Life Sciences document (Department of Education [DoE] 2011). The third aim is to enhance learners' understanding of the relevance of school science to their lives, and how science can enrich their lives. It promotes the teaching of science in an integrated way to enhance the "relationship between the (science) subject and society" (DoE 2011, p.17). Certain social challenges with substantive connections to science, for example, HIV/AIDS, the green economy, indigenous knowledge systems and sustainability (Lotz-Sisitka et al. 2015; Mnguni 2017; Mudaly 2011) have been explored among South African scholars. Mnguni (2017), who analysed the curriculum to explore the infusion of SSIs in the curriculum, found that the integration of two SSIs, namely, the green economy and HIV/AIDS, was poor.

Many studies on SSI focus on teachers' motivation, views and beliefs about incorporating SSI (Lee and Witz 2009). There is a dearth of studies which explore practices of teachers who do incorporate SSI into their curricula (Klosterman et al. 2012). This study responds to this paucity in research by exploring the integration of SSIs by three novice, practicing Life Sciences teachers. The following question is central to this study: *How do novice Life Sciences teachers integrate SSIs into the curriculum, within the context of decolonising the curriculum?* Critical theory, and more specifically, critical pedagogy, framed the study.

11.3 Critical Pedagogy

Critical theory contests the existing state of affairs in contemporary (science) education (Weston 2015). Science teacher pedagogy is based on external interactions, which include what teachers understand through their interaction with the curriculum, their professional development, and requirements based on curriculum policy. However, teachers can produce knowledge differently, by being agents of self-production, and this involves internal interaction in teacher pedagogy (Degener 2001; Magill and Rodriguez 2014; Oestereicher 1979). Although the curriculum, which is imbued with innocuous forms of oppression, is "reproductive and often culturally disparaging", this can be changed by teachers "asserting their self-determination" (Magill and Rodriguez 2014, p. 220). A critical pedagogue considers different socially constructed forces, such as racism, ethocentricism and classism, used to subordinate learners (Degener 2001). According to Giroux (2004), critical pedagogy is crucial in the struggle for democracy. Critical pedagogy underscores the role of critical reflexivity, creates connections between what is taught and daily life experiences, provides an understanding of the interrelatedness between knowledge and power, and uses resources in history to advance people's democratic rights (Giroux 2004). Critical pedagogy involves teaching which considers the backgrounds and worldviews of learners, and transforms the school and curriculum into a democratic site, and teachers and learners as agents of change (Giroux 2011).

11.4 Teachers, Learners and the Curriculum

Advocates of critical pedagogy express concern about how teachers become enslaved in a neoliberalist capitalist framework by education departments, which prescribe textbooks and materials which normalise dominant (Euro-Western) values (Maistry 2015; Magill and Rodriguez 2014). Lacan (2006), cited by Magill and Rodriguez (2014, p.215), challenges teachers not to accept the curriculum as "objet petit a", that is, with passivity, devoid of internal consciousness, about the what, how and why of teaching. Magill and Rodriguez (2014) assert that the most potent in-school factor which influences learners is an effective teacher. Teachers should be critical of a curriculum which is underpinned by dominant ideals of particular groups of people. This type of curriculum reduces learners to "cheerful robots" by adopting pedagogies which claim to be value-free and eschews issues related to epistemic justice, knowledge and power, and ethics (Giroux 2011, p. 3).

Learners encounter "null spaces" which are what transpires between the planned and experienced curriculum, and these mould learners to internalise and accept a curriculum of domination and their subaltern status in society. Ultimately, learners become compliers of ideologies which they had imbibed through the curriculum (Heleta 2016; Motta 2013). Motta (2013) adds that learners experience forced enculturation and silencing of alternative knowledge systems and discourses when curricula reproduce epistemological blindnesses. These curricula perpetuate social stratification by excluding learners who are in "subordinate" race, gender, etholinguistic, and cultural settings (Heleta 2016; Motta 2013). Teachers can position themselves as key actors in a curriculum for human agency which can "relate to diverse cultural needs within a pluralistic society" (Magill and Rodriguez 2014, p. 218).

11.5 Methodology

A qualitative, case study methodology was adopted to explore novice Life Sciences teachers' preparation for and implementation of SSI-based instruction. The methodology was appropriate because a variety of methods were adopted to achieve a deep understanding (Cohen et al. 2011) of how teachers prepared for and taught SSI-based lessons. An illustrative case study was central to this design. Yin (2009, p.18) stated that "a case study is an empirical inquiry that investigates a contemporary phenomenon in-depth and within its real-life context". For Hayes et al. (2015), an illustrative case study involves a detailed description of and reasons for activities and events related to a phenomenon (in this study, the phenomenon of how teachers prepare for and implement science lessons which are connected to social issues). The study involved activities in an Honours module which encouraged teachers to transcend boundaries of familiar ways of preparing for and teaching science lessons. The study was embedded in the critical philosophical paradigm, and encouraged teachers to become critical social agents (le Grange 2016; Magill and Rodriguez 2014) by entering a transdisciplinary space, which focussed on SSIs within the context of decolonising the curriculum.

Data was generated from portfolios of evidence and reflective journals. Singh et al. (2015) emphasise the suitability of portfolios in capturing evidence for tasks which are process-orientated. In this study, teachers recorded their preparation for and delivery of lesson in their portfolios. According to Tillema et al. (2011) reflective journals are useful because they provide insight into the cyclical process of professional growth. Themes were derived from the data set using content analysis. The credibility and accuracy of the data was ensured by triangulating data from multiple sources (Creswell 2012).

11.5.1 The Participants

The sample comprised three purposively selected practicing, fully qualified teachers. The criteria for their selection was that they taught Life Sciences, had less than 5 years of teaching experience, and were registered to study an Honours in Science and Mathematics Education module in curriculum development. A brief biography of the participants is provided in Table 11.1.

The participants were novice, practicing teachers who taught Natural Sciences (Grades 8 and 9) or Life Sciences (Grades 10, 11 and 12). Each teacher taught in a different school in an urban area in the province of Kwa-Zulu Natal. The average number of learners in the Life Sciences classroom was 35. Each school was under-resourced and none of these teachers had access to a science laboratory. They relied on innovating and improvisation to conduct practical work.

Table 11.1 Biography of participants

	Teacher A	Teacher B	Teacher C
Gender	Male	Female	Female
Number of years of experience	3	4	1
Subjects taught	Life Sciences	Life Sciences	Life Sciences
	English	Natural Sciences	Natural Sciences
Qualifications	Bachelor of Education (Majors: Biological Science, English)	Bachelor of Education (Majors: Biological Science, Sports Science)	BSc Environmental Science (Majors: Plant studies, Animal studies, Ecology and Resource management. Postgraduate Certificate in Education
	Bachelor of Education (Honours in Science and Mathematics Education)		
	Currently enrolled for Masters in Science Education		

11.5.1.1 The Task

Bearing in mind the perspective of Giroux (2011, p. 3) that critical pedagogy "also provides tools to unsettle common sense assumptions, theorize matters of self and social agency", and in an effort to integrate the call for decolonising the curriculum into the Honours in Science and Mathematics Education module, the following activity was developed:

> Preamble: *The Curriculum is viewed as a vehicle to achieve the aims of the Constitution, and, to this effect, the Curriculum Assessment Policy Statement (CAPS) includes the following principles: "Social transformation: ensuring that the educational imbalances of the past are redressed, and that equal educational opportunities are provided for all sections of the population" and "Human rights, inclusivity, environmental and social justice: infusing the principles and practices of social and environmental justice and human rights as defined in the Constitution of the Republic of South Africa. The National Curriculum Statement Grades R-12 is sensitive to issues of diversity such as poverty, inequality, race, gender, language, age, disability and other factors" (DoE 2011, pp.4-5). However, analysts argue that the curriculum continues to be embedded in colonial, apartheid and western worldviews.*

Task

- Critique a unit of work in the science curriculum to determine whether the social justice and social transformation ideology (as per policy document in the preceding preamble) underpin the work, given the call to decolonise the curriculum.
- Re-design the unit of work to address a social issue/challenge through the science lessons, and which addresses social transformation or social justice.
- Reflect on these activities and record your reflections in a journal.

The purpose of the task was to motivate teachers to consider the ethical implications of what and how they teach. Teachers were encouraged to design units of work in which science issues were linked to challenges encountered by the community. This was done to enhance teachers' "reflective judgement" and create meaningful, relevant science lessons (Burek and Zeidler 2015, p.425).

11.6 Results

The results are presented in three parts for each teacher, and comprised teachers' responses to the following questions:

- Why did you choose (topic) as an SSI which could decolonise the unit of work?
- How did you prepare to teach this SSI?
- Describe the lesson plan you designed which informed your teaching of the SSI.

Teachers A and B selected "Treatments for Cancer" as the topic. This is a subtopic of the Cell Cycle in the Knowledge Strand titled *Life at the Molecular, Cellular and Tissue Level* (DoE 2011). According to the CAPS document, learners are required to research and write a report on causes, beliefs and attitudes and treatment (radiography, chemotherapy) on one type of cancer (DoE 2011, p. 26). Teacher C selected "Alien Invasive Plants" as a topic in the Knowledge Strand titled *Environmental Studies*. This is a minor part of the larger topic: "Human Impact on the Environment: Current Crises for Human Survival." Learners are expected to select an alien plant and write a report on it, using text-based resources (textbooks, reference books, ShareNet, reports in the media) (DoE, 2011, p. 51). The data is presented using following codes to describe data sources: Reflective Journal – (RJ), and Portfolio – (P).

11.6.1 Results from Teacher A

11.6.1.1 Reasons for Selecting Alternative Treatments for Cancer as an SSI

> "There were some incidents of cancer in the community. But being a historically indigenous context many illnesses were dealt with mainly using indigenous ways. This served as a way to raise learners' awareness of other treatments that do exist. IK was a starting point." (RJ)

11.6.1.2 Preparing to Teach Alternative Treatments for Cancer as an SSI

The following entries were made in the reflective journal about how Teacher A prepared to teach the SSI.

> "I analysed Curriculum (CAPS) documents. The CAPS document highlights only two forms of treatment for cancer which focuses on medical biotechnology. They are Radiotherapy and Chemotherapy. Both these treatment forms have western origins.
>
> This means that Westerners are seen as the only people with enough powerful knowledge on cancer to propose treatment while the rest of the world buys into this. Western medical industries make vast sums of money from cancer treatment. Other types of medical

options are seen as useless fads. Learners will be colonised to believe that only Western forms of treatment are useful. Since only Western treatment is acknowledged in the curriculum, non-western South African learners may feel marginalised and inferior. People will not have access to expensive Western cancer treatments and they could die because of this." (RJ)

The following motivation for the selection of the topic was documented in the portfolio.

"I did a survey of learners' beliefs and views about cancer to understand learners' prior knowledge and beliefs about cancer in their communities. Then it was (my) reading articles and research from literature mostly. But other experienced teachers who were in tune with IK offered guidance as well. The lessons I prepared included African indigenous treatments in addition to Western treatments." (P)

Summary of lesson plan which informed teaching about alternative treatments for cancer

Grade: 10		
Knowledge Strand: Life at the molecular, cellular and tissue level		
Topic: Cancer treatments		
Aim:		
access information about cancer treatments from a variety of sources (reference books, textbooks, internet, community elders)		
critically evaluate scientific information and functionality of different cancer treatments		
appreciate different world views in treatment of cancer		
Resources: projector for powerpoint presentation,		
Teacher activities	Learner activities	Teaching strategy
Present powerpoint slides on alternative treatments for cancer	Listen to presentation, contribute to discussion	Teacher-learner discussion
Facilitate group work	Draw on previous homework task about common treatments for cancer from alternative knowledge systems. Use electronic devices (tablets, cell phones, laptops), articles from scholarly texts (research periodicals) to search for information. Discuss and debate your views about the usefulness of alternative treatments for cancer. Create a poster to report findings	Group work, discussion
Facilitate class discussion	Poster presentation by learners	Class discussion

Source: Portfolio of Teacher A

Slides were included in the powerpoint presentation, and contained pictures and summaries of treatment for cancer based on naturopathy, homeopathy and indigenous knowledge.

11.6.2 Results from Teacher B

11.6.2.1 Reasons for Selecting African Treatments for Cancer as an SSI

The following extract was sourced from the portfolio:

> "Information regarding traditional African treatments for cancer in the current curriculum is very brief and through decolonisation of the information it would lead to a deeper understanding of where this information came from." (P)

In her reflective journal, Teacher B wrote:

> "Cancer seems to be a disease with in the Pietermaritzburg area that everyone knows about because they have a friend or family member who has been affected by it or is currently suffering from it. On a more personal note I have seen how people who undergo cancer treatments which are westernised (chemo, surgery, radiotherapy) often suffer from extreme side effects and feel that personally there needs to be research into alternative treatments that have less server side effects. And yes a large part of this was to draw learners' attention to what was not in the textbooks and to give recognition to practices that occur around them and to show them that traditional African ways are valued." (RJ)

11.6.2.2 Preparing to Teach about African Treatments for Cancer as an SSI

In her reflective journal, Teacher B made the following entry:

> "All information that I used to plan a decolonised lesson was from the Internet … there are articles of indigenous and other knowledges to manage cancer." (RJ)

In the portfolio, Teacher B described how she prepared her lesson:

"I studied the CAPS curriculum and textbooks. In analysing the prescribed textbook I found that:

- Treatment options listed in textbooks are of Western origin: Chemotherapy, Radiotherapy, Surgery
- Availability of the best treatments depends on social class. Those who can afford it receive best treatment.
- Scientists shown are of European or North American descent. Textbook idolises European scientists and leaves no hope for the African child who.....might believe that scientific inquiry and discovery is not for the African child because African are not depicted in the everyday teaching of science." (P)

Teacher B provided photographic evidence (in Fig. 11.1) to show that textbooks presented information which privileged Euro-Western knowledge.

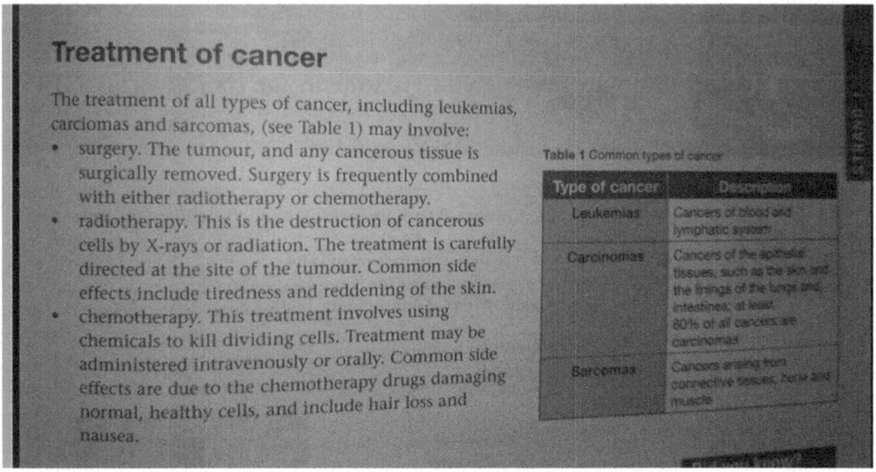

Treatment of cancer

The treatment of all types of cancer, including leukemias, carciomas and sarcomas, (see Table 1) may involve:

- surgery. The tumour, and any cancerous tissue is surgically removed. Surgery is frequently combined with either radiotherapy or chemotherapy.
- radiotherapy. This is the destruction of cancerous cells by X-rays or radiation. The treatment is carefully directed at the site of the tumour. Common side effects include tiredness and reddening of the skin.
- chemotherapy. This treatment involves using chemicals to kill dividing cells. Treatment may be administered intravenously or orally. Common side effects are due to the chemotherapy drugs damaging normal, healthy cells, and include hair loss and nausea.

Table 1 Common types of cancer

Type of cancer	Description
Leukemias	Cancers of blood and lymphatic system
Carcinomas	Cancers of the epithelial tissues, such as the skin and the linings of the lungs and intestines; at least 80% of all cancers are carcinomas
Sarcomas	Cancers arising from connective tissues, bone and muscle

Fig. 11.1 Section of a prescribed science textbook to show privileging of Euro-Western knowledge

Summary of lesson plan which informed teaching about alternative treatments for cancer

Grade: 10		
Knowledge Strand: Life at the molecular, cellular and tissue level		
Topic: Cancer treatments		
Aim:		
access information about cancer treatments from a variety of sources (reference books, textbooks, internet, community elders)		
critically evaluate scientific information about who researches cancer treatment and what treatment options are presented		
relate AIK (African Indigenous Knowledge) to cancer and show its application in daily life		
Resources: projector for powerpoint presentation, electronic devices		
Teacher activities	Learner activities	Teaching strategy
Present powerpoint slides on treatments for cancer and scientists involved	Listen to presentation, contribute to discussion	Teacher-learner discussion
Facilitate group work	Refer to previous homework task about common treatments for cancer. Use laptops, tablets, cell phones, to search for information on treatments from African indigenous knowledge systems. Argue about the pros and cons of using indigenous knowledge treatments for cancer. Complile a report.	Group work, discussion
Facilitate class discussion	Report back by learners	Class discussion

Source: Portfolio of Teacher B

Fig. 11.2 Female African knowledge producer in cancer research

In the powerpoint presentation a slide on the Kraalbos plant, among others, was included to show how Africa Indigenous plants can be used to treat the disease. A South African scientist, Professor Burtram Fielding, who found chemical compounds from the African indigenous Kraalbos plant to be effective against breast cancer cells, was also included among the slides.

Another African researcher in the field of cancer research was presented in a slide. A picture of Professor Tebello Nyokong, who is a South African chemist at Rhodes University was included (Fig. 11.2). Her research on photo-dynamic therapy as an alternative treatment for cancer was presented (P):

11.6.3 Teacher C

11.6.3.1 Reason for Selecting Alien Invasive Plants as an SSI

Teacher C indicated in her reflective journal that the school environment was the reason for selecting the topic.

> "The school yard is overgrown with alien invasive plants. I thought 'Of what real use is this (writing a report on an alien invasive plant), when the whole area, not just the school, is full of them (alien plants)?'" (RJ)

11.6.3.2 Preparing to Teach About Alien Invasive Plants as an SSI

Teacher C presented the following insights into how she prepared SSI-based instruction in her portfolio.

> "I saw the CAPS (read curriculum documents) and this topic is done under Human Impact on the Environment, for the purpose of environment management. For practicals learners must observe the impact of alien species on biodiversity, according to CAPS. CAPS wants

learners to write a report on an example of an alien invasive plant. This is not a good way to manage the invasion of alien plants by science learners. I wanted them (learners) to see these plants in their environment, know how these plants affect the environment negatively, and to remove them. I felt that by increasing learners' knowledge about these plants and letting them do a hands-on activity to remove these plants would be more relevant. Also I have a degree in Nature Conservation and my major subjects are Plant studies, Animal studies, Ecology and Resource management. I felt I could handle the topic nicely." (P)

"I took photographs of alien plants in the school yard. I identified the plants and read about how they can be eradicated. I used information from textbooks and charts and also worked with Environmental Officer from Natural Resources. I studied the textbook and other texts about alien vegetation management." (P)

Summary of lesson plan which informed teaching about alien invasive plants

Grade: 11		
Knowledge Strand: Environmental studies		
Topic: Human impact on environment and environmental management of alien plants		
Aim:		
Describe how alien vegetation can be managed using biological, manual, chemical and mechanical methods.		
Describe the effects of alien plants on		
Water resources		
Agricultural production		
Biodiversity		
Identify alien plants in the school yard and remove these by mechanical means		
Resources: Textbook, video, alien invasive species regulations policy document,		
Teacher activities	Learner activities	Teaching strategy
Video presentation to show effects of alien invasive plants	Watch video and discuss	Teacher-learner discussion
Distribute handouts with pictures and names of common alien invasive plants, and mechanical removal. Facilitate group discussion	Study handouts in groups	Group work, discussion
Facilitate field work	Work in groups of 5 or 6. Go into the school yard and identify alien invasive plants. Uproot smaller plants. Use a slasher to remove larger plants under teacher supervision. Use a hoe to remove root balls or tap roots.	Field work

Source: Portfolio of Teacher C

11.7 Analysis and Discussion

Borrowing from Owens et al. (2017), the phases of the SSI-instruction framework, which were adapted in this study to apply to the teacher (instead of student) knowledge and practices, informed the analysis and findings. Common themes were generated from the results sourced from Teachers A, B and C. These themes are *identifying the social challenge, preparing a science lesson which infuses the issue,* and *addressing the issue through teaching.*

11.7.1 Identifying the SSI

Teacher A had knowledge of cancer treatment used in by the community members. He was aware that many members of the community relied on IK to manage the disease. The affordability of treatments by middle class people was also mentioned by Teacher A. In order to access learners' views and understanding of cancer, Teacher A used a survey. Teacher A did not assume to know learners' knowledge and experiences, and this was the reason for conducting a survey. Le Grange (2016, p.9) highlighted the importance of creating spaces for "voices and knowledges" of marginalised people.

Teacher B had knowledge of the incidence of cancer in the community, when she stated "cancer seems to be a disease with in the Pietermaritzburg area that everyone knows about because they have a friend or family member who has been affected by it or is currently suffering from it." In addition, she was aware of the dangerous side effects of Western methods of treatment, and this is evident by her statement: "I have seen how people who undergo cancer treatments which are westernised (chemo, surgery, radiotherapy) often suffer from extreme side effects and feel that personally there needs to be research into alternative treatments that have less severe side effects."

Teacher C was acutely aware of the invasion of alien vegetation in the school grounds. She was also aware of the invasion of alien plants in the wider community, when she said: "the whole area, not just the school, is full of them (alien plants)." Her awareness was possibly heightened due to her expert knowledge on the phenomenon, having qualified with a BSc in Environmental Science. She used this as an opportunity to teach about alien invasive plants as an SSI.

All three teachers identified the SSI based on their knowledge of the context, specifically that of the community. Teacher A had knowledge of community members' response to disease based on their worldviews which were dissimilar to that of Western science. Teacher B had knowledge of incidence of cancer in the community. In addition, she had concerns about the side effects of conventional Western treatments for cancer. This resonates with the assertion by Santos (2009) that society be made aware of both the benefits and risks associated with products of modern science. Teacher C was motivated to incorporate alien invasive plants into the teaching based on her knowledge of the flora of the community. Unlike teachers A and B whose motivation was based on community members' practices, Teacher C's motivation was based on environmental issues in the community.

11.7.2 Preparing to Teach SSI: Deconstructing the Curriculum

Teacher A enhanced his knowledge of the curriculum by analysing it using a critical lens. He examined both the explicit and the hidden curriculum. His analysis of the curriculum revealed that it ignored the treatment used by many community members. In this way, the curriculum could have marginalised learners, by not addressing what they were familiar with in daily life, according to Teacher A, who stated "Since only Western treatment is acknowledged in the curriculum, non-western South African learners may feel marginalised and inferior." He concluded that the curriculum valued Euro-Western epistemic frameworks because knowledge producers (scientists of Euro-Western descent), the context in which knowledge was produced (North America and Europe) and the type of knowledge (radiotherapy and chemotherapy as treatments for cancer) in the curriculum privileged Euro-Western knowledge traditions. The complicity of the curriculum in neglecting non-Western worldviews and knowledge systems is well documented (Heleta 2016; le Grange 2016).

Teacher A enhanced his subject matter knowledge by reading texts about alternative treatments for cancer. He networked with more experienced teachers to improve his capacity to teach. The advantage of appropriating epistemic knowledge by interacting with more experienced colleagues was advocated by Maistry (2015) and Mudaly (2015).

Teacher B analysed the curriculum critically. In addition, she analysed prescribed textbooks and found that the content and underlying ideology privileged Euro-Western paradigms, when she stated: "Treatment options listed in textbooks are of Western origin…, Scientists shown are of European or North American descent. Textbook idolises European scientists and leaves no hope for the African child." For Teacher B, the lack of role models with which the African learner could identify, impacted negatively on the learner. The insidious denigration and subjugation of people, by using the institution of education, was emphasised by le Grange (2018) and Connell (2016).

Teacher C conducted an analysis of the curriculum documents. She found that the response to alien invasive plants, as outlined in the curriculum, lacked meaning for learners. She asserted: "CAPS (curriculum) wants learners to write a report on an example of an alien invasive plants. This is not a good way to manage the invasion of alien plants by science learners. I wanted them (learners) to see these plants in their environment, know how these plants affect the environment negatively, and to remove them…" The importance of science education which is relevant to learners has been underscored in several studies (Holbrook and Rannikmae 2007; Mudaly 2011). Teacher C prepared to teach using visual methods, by taking photographs of alien plants in the school yard and identifying them, using reference books. She also leveraged support from an external expert, by working with the "Environmental Officer from Natural Resources." Obtaining outside support for developing new teacher knowledge was recommended by Rogan (2007).

11.7.3 Addressing the Social Challenge through SSI-Based Instruction: Reconstruction

Teacher A re-imagined the curriculum by adding other remedies for the treatment of cancer in his presentation of powerpoint slides. Learners were encouraged to take responsibility for their learning by doing a homework exercise on cancer treatments from alternative knowledge sources. Learners were encouraged to engage in "recovering and re-discovering" (le Grange 2016, p.3) knowledge about treatment of cancer. This disrupted the dominance of treatments from Euro-Western sources, and learners were enabled to access information in groups about the use of alternative (to chemotherapy and radiotherapy) treatments, using electronic devices and scholarly texts. Learner centred pedagogy informed the lesson, where learners were encouraged to debate usefulness of alternative treatments, based on the information they sourced. Alternative knowledge which had been excluded during centuries of oppression (Kruger and Fataar 2017), was (re)appropriated by learners. Critical thinking was enhanced by asking learners to "discuss and debate your views about the usefulness of alternative treatments." The importance of debates in reconstructing the curriculum was emphasised by Santos (2009).

Teacher B expanded her teaching to transcend what was stipulated in prescribed textbooks and the curriculum, by including African indigenous plants and a female, African scientist who researched alternative therapy for cancer, to learners. This marked a departure from the monocultural perspectives (Kincheloe 2008) of who is a legitimate knowledge holder, and what is valuable knowledge. It resonated with Hountonji's call for African knowledge to be reappropriated and for "épistémologies du Sud" to be revived (Hountonji 2009, p.1). The importance of female role models in science education was underscored by Juan et al. (2018).

In Teacher B's classroom, a learner centred pedagogy was adopted and learners worked collaboratively in groups to actively construct knowledge about the use of African indigenous plants in treating cancer. Learners were encouraged to use argumentation to think critically as they constructed knowledge. Insight into the value of argumentation as a strategy to provoke thoughts about controversial SSIs was provided in another study by Karahan and Roehrig (2019).

Teacher C used a video as an aid to teach about the effects of alien plants on the environment. She worked with an external expert, an Environmental Officer to develop worksheets to enable learners to identify alien plants. Under her supervision, learners identified and eradicated alien plants in the school yard mechanically. This addresses the plea from Hountonji (2009) for the production of knowledge which meets the needs of African societies. Teacher C's engagement of learners in field work transcended the normal boundaries of activities prescribed in the curriculum. She developed in the learners the skill of identifying and removing alien invasive plants from the school, instead of following the curriculum guideline for learners to write a report on an alien plant.

The three teachers in this study adopted different strategies to implement lessons which were embedded with SSIs. They developed learner-centred lessons and encouraged group work and peer collaboration. Teachers A and B shifted the responsibility of accessing information to the learners. They were encouraged to think critically about the information which they collated and presented as posters or reports. Argumentation and debating were encouraged as learners constructed knowledge collaboratively. Teacher C prepared the worksheets to guide learners to identify alien plants, and this was dissimilar to the strategy used by Teachers A and B. However, Teacher C adopted a practical approach by involving learners in fieldwork. Teacher C was the only participant to engage learners in hands-on activities. Teachers A and B privileged minds-on activities.

11.8 Concluding Remarks

Teachers' professional knowledge and experiences informed ways in which they integrated SSIs in the curriculum within the context of decolonising the curriculum. Their reasons for selecting specific topics was based on their knowledge of the context, particularly that of the community. Teachers A and B understood the incidence of cancer in the community. Teacher A also demonstrated knowledge of treatments which community members preferred. Teacher C, a graduate in environmental science, was possibly more aware of alien invasive plants, based on her professional knowledge. This, combined with the invasion of alien plants in the school and community, spurred her to respond within the context of science lessons.

In preparing to teach lessons which integrated socioscientific issues, the teachers, as critical pedagogues, analysed curriculum documents. This involved deconstructing the unit of work which was outlined in the curriculum and in textbooks. Analyses by Teachers A and B rendered elements of the hidden curriculum more visible. The promotion of Euro-Western knowledge and knowledge producers in the curriculum and textbooks was perceived as a form of epistemic injustice. The critique of the curriculum by Teacher C revealed that the information on alien plants and the learner activities stipulated in the curriculum were inadequate and irrelevant.

Teacher A enhanced his subject matter knowledge and pedagogical content knowledge by collaborating with more experienced teachers. Teacher C leveraged support from an Environmental Officer, an external expert, to deepen her subject matter knowledge. Each participant used the internet, textbooks and other reference materials to deepen their knowledge of science which was related to the topic.

In order to address the social challenge through SSI-based teaching, teachers reconstructed the units of work. Teachers A and B disrupted the process of re-inscribing dominant positions of Western figures by including non-Western scientists in their teaching. Teachers A and B included opportunities for learners to demystify IKS by engaging them in activities where they were required to actively access information. Pedagogic activities were designed to stimulate critical thinking and argumentation among learners. Teacher C engaged learners in a field study, where they eradicated alien plants, and contributed positively to the school environment. Each teacher was intellectually imaginative and cast learners as autonomous, self-reliant individuals.

This study revealed ways in which teachers adopted a human-centric approach to teaching, in order to address challenges in their communities. As critical social agents, they developed units of work which were relevant to the school communities. They asserted their self-determination by adopting an alternative pedagogic vision when designing lessons which valued lived experiences of learners and their communities.

References

Burek, K., & Zeidler, D. L. (2015). Seeing the forest for the trees! Conservation and activism through socioscientific issues. In M. P. Mueller & D. J. Tippins (Eds.), *EcoJustice, citizen science and youth activism: Situated tensions for science education* (pp. 425–442). Dordrecht: Springer International Press.

Cohen, L., Manion, L., & Morrison, K. (2011). *Research methods in education* (7th ed.). London: Routledge.

Connell, R. (2016). *Decolonising knowledge, democratising curriculum*. Retrieved from https://www.uj.ac.za/faculties/humanities/sociology/PublishingImages/Pages/Seminar/Raewyn%20Connell's%20Paper%20on%20Decolonisation%20of%20Knowledge.pdf

Creswell, J. W. (2012). *Educational research: Planning, conducting, and evaluating quantitative and qualitative research* (4th ed.). Boston: Pearson.

Degener, S. C. (2001). Making sense of critical pedagogy in adult literacy education. *Annual Review of Adult Learning and Literacy, 2*(1), 26–62.

Department of Basic Education (DBE). (2011). *National Curriculum Statement (NCS): Curriculum and Assessment Policy Statement (CAPS)*. Further Education and Training Phase: Grades 10–12. Pretoria: Government Printer.

Giroux, H. A. (2004). Critical pedagogy and the postmodern/modern divide: Towards a pedagogy of democratization. *Teacher Education Quarterly, 31*(1), 31–47.

Giroux, H. (2011). *On critical pedagogy*. New York: Bloomsbury.

Hayes, R., Kyer, B. & Weber, E. (2015). *A case study cookbook*. Retrieved from https://web.wpi.edu/Pubs/E-project/Available/E-project-121615-164731/unrestricted/USPTO_CookbookFinal.pdf

Heleta, S. (2016). Decolonization of higher education: Dismantling epistemic violence and Eurocentrism in South Africa. *Transformation in Higher Education, 1*(1), 1–8.

Holbrook, J., & Rannikmae, M. (2007). The nature of science for enhancing scientific literacy. *International Journal of Science Education, 29*, 1347–1362.

Hountonji, P. J. (2009). Knowledge of Africa, Knowledge by Africans: Two perspectives on African studies. *RCCS Annual Review, 1*. Retrieved from http://www.ces.uc.pt/publicacoes/annualreview/media/2009%20issue%20n.%201/AR1_6.PHountondji_RCCS80.pdf

Juan, A., Hannan, S. & Namome, C. (2018). I believe I can do science: Self-efficacy and science achievement of Grade 9 students in South Africa. *South African Journal of Science, 114*(7/8), Art. #2017-0269, 7 pages. https://doi.org/10.17159/sajs.2018/20170269. Retrieved from http://www.scielo.org.za/pdf/sajs/v114n7-8/15.pdf

Karahan, E., & Roehrig, G. H. (2019). Case studies of science teachers designing socioscientific issues-based instruction. *Hacettepe University Journal of Education., 34*(1), 71–89. https://doi.org/10.16986/HUJE.2018044772.

Kincheloe, J. L. (2008). *Critical pedagogy of the knowledge of wars of the twentieth century*. Retrieved from http://friereproject.org/wp-content/journals/TIJCP/Vol1No1/48-38-1-PB.pdf

Klosterman, M. L., Sadler, T. D., & Brown, J. (2012). Science teachers use of mass media to address socio-scientific and sustainability issues. *Research in Science Education, 42*(1), 51–74.

Kruger, E., & Fataar, A. (2017). *Decolonising education in South Africa: An interview with Aslam Fataar*. University Seminar, 23 January. Retrieved from https://www.litnet.co.za/decolonising-education-south-africa-interview-aslam-fataar/

Lacan, J. (2006). *Ecrits*. New York: W. W. Norton & Co.

le Grange, L. (2016). Decolonising the university curriculum. *South African Journal of Higher Education, 30*(2), 1–12.

le Grange, L. (2018). Decolonising, Africanising, indigenising, and internationalising curriculum studies: Opportunities to (re)imagine the field. *Journal of Education*, (74), 4–18.

Lee, H., & Witz, K. G. (2009). Science teachers' inspiration for teaching socio-scientific issues: Disconnection with reform efforts. *International Journal of Science Education, 31*(7), 931–960.

Lotz-Sisitka, H., Wals, A. E. J., Kronlid, D., & McGarry, D. (2015). Transformative, transgressive social learning: rethinking higher education pedagogy in times of systemic global dysfunction. *Current Opinion in Environmental Sustainability, 16*, 73–80.

Magill, K., & Rodriguez, A. (2014). *A critical humanist curriculum*. Retrieved from https://scholarworks.boisestate.edu/cgi/viewcontent.cgi?article=1028&context=esl_facpubs

Maistry, S. (2015). Towards a humanising pedagogy: An autoethnographic reflection of my emerging postgraduate research supervision practice. *Journal of Education, 62*, 60–85.

Mnguni, L. (2017). Curriculum ideologies and socioscientific issues in life sciences. In *Proceedings of the ISTE international conference on mathematics, science and technology education*, 2016, 23–26 October 2017, Kruger National Park, Mopani camp, Phalaborwa, Limpopo, South Africa (pp. 127–136). ISBN: 978-1-86888-983-9.

Motta, S. C. (2013). Teaching global and social justice as transgressive spaces of possibility. *Antipode, 45*(1), 80–100. Retrieved from https://onlinelibrary.wiley.com/doi/pdf/10.1111/j.1467-8330.2012.00995.x.

Mudaly, R. (2011). Risking it: Entering uneven socio-scientific spaces in a life sciences classroom. *African Journal of Research in Mathematics, Science and Technology Education (special issue), 15*(3), 27–40.

Mudaly, R. (2015). Creating my academic self and space: Autoethnographic reflections on transcending barriers in higher education. *Journal of Education, 62*, 35–58.

Oestereicher, E. (1979). The self-production of society. *Social Forces, 57*(3), 1006–1007.

Onwu, G. O. M., & Kyle, W. C., Jr. (2011). Increasing the socio-cultural relevance of science education for sustainable development. *African Journal of Research in Mathematics, Science and Technology Education, 15*(3), 5–26.

Osbourne, J., Duschl, R., & Fairbrother, R. (2002). *Breaking the mould: Teaching science for public understanding*. Retrieved from https://www.nuffieldfoundation.org/sites/default/files/Breaking%20the%20mould_science%20for%20public%20understanding.pdf

Owens, D. C., Sadler, T. D., & Zeidler, D. L. (2017). Controversial issues in the science classroom. *Phi Delta Kappan, 99*(4), 45–49.

Pinar, W. F. (2004). "A lingering note": An introduction to the collected works of Ted T. Aoki. In W. F. Pinar & R. L. Irwin (Eds.), *Curriculum in a new key: The collected works of Ted T. Aoki* (pp. 1–85). Lawrence Erlbaum Associates: Mahwah.

Rogan, J. M. (2007). How much curriculum change is appropriate? Defining a zone of feasible innovation. *Science Education, 91*(3), 439–460.

Santos, W. L. P. D. (2009). *Scientific literacy: A freirean perspective as a radical view of humanistic science education*. https://doi.org/10.1002/sce.20301.

Singh, C. K. S., Samad, A. A., Hussin, H., & Sulaiman, T. (2015). Developing a portfolio assessment model for the teaching and learning of English in Malaysian L2 classroom. *English Language Teaching, 8*(7), 164–173

Talens, J. (2016). Teaching with socio-scientific issues in physical science: Teacher and students' experiences. *International Journal of Evaluation and Research in Education (IJERE), 5*(4), 271–283.

Tillema, H. H., Smith, K., & Leshem, S. (2011). Dual roles–conflicting purposes: A comparative study on perceptions on assessment in mentoring relations during practicum. *European Journal of Teacher Education, 34*(2), 139–159.

Weston, D. (2015). *The political economy of global warming: The terminal crisis*. London: Routledge.

Yin, R. K. (2009). *Case study research: Design and methods* (4th ed.). Los Angeles: Sage.

Zeidler, D. L., & Nichols, B. H. (2009). Socioscientific issues: Theory and practice. *Journal of Elementary Science Education, 21*(2), 49–58.

Ronicka Mudaly is an associate professor in the School of Education at the University of KwaZulu-Natal in South Africa. A quest for a science education which is socially just and more relevant, inform her pedagogy. Her current research interests include humanist approaches in science education from a Freirean perspective, indigenizing education, gender studies and youth empowerment through research.

Chapter 12
New Perspectives for Addressing Socioscientific Issues in Teacher Education

Jan Alexis Nielsen, Maria Evagorou, and Justin Dillon

12.1 Introduction

The purpose of this book was to bring together international researchers working on teacher professional development, with an emphasis on SSI, to share their work. We are proud to see the truly international and multifarious character of the preceding 11 chapters. As such, the chapters reflect just how diverse the international landscape of science teaching, in general, and of SSI teaching, in particular. We want to use this final chapter to connect some of the larger threads that seem to run across multiple chapters. We identify three main emergent themes:

- Teachers' (and student teachers') backgrounds and beliefs are often deciding factors in the uptake and quality of SSI teaching,
- SSI teaching is often not the sole "new" pedagogical principle that is being implemented – e.g. SSI will often be combined with inquiry – and that has benefits and challenges,
- (Long-term) professional development or training of student teachers is needed in order to facility the uptake and quality of SSI teaching.

After presenting these themes through the findings of the individual chapters, we identify gaps in knowledge that still need to be covered even after this volume.

J. A. Nielsen (✉)
Department of Science Education, University of Copenhagen, Copenhagen, Denmark
e-mail: janielsen@ind.ku.dk

M. Evagorou
Department of Education, University of Nicosia, Nicosia, Cyprus
e-mail: evagorou.m@unic.ac.cy

J. Dillon
Graduate School of Education, Exeter University, Exeter, UK
e-mail: J.S.Dillon@exeter.ac.uk

M. Evagorou et al. (eds.), *Science Teacher Education for Responsible Citizenship*, Contemporary Trends and Issues in Science Education 52,
https://doi.org/10.1007/978-3-030-40229-7_12

12.2 Teachers' Background and Beliefs

The first emergent theme from the contributions in this volume relates to how the background of teachers and student teachers – in terms of their knowledge, skills and attitudes – mediates the implementation of SSI-teaching. In Chap. 3, Leung et al. (2020) present an intricate connectedness of the beliefs about and priority of SSI and a range of other components of the curriculum in Hong Kong. In particular, they found that the reason that some pre-service teachers seem to devalue SSI teaching is a type of (mistaken) belief about how content knowledge, nature of science can relate to SSI – namely that the only relation between the three components is that SSI is a vehicle for teaching the other two. This resonates with the findings about in-service teachers by Tidemand and Nielsen (2017) that SSI is often reduced to an instrument for teaching other parts of the curriculum – most notably content knowledge.

In Chap. 10, Garrido Espeja and Couso (2020) found that, for pre-service teachers, the most challenging aspect about teaching SSI activities was to include the scientific information in a way that facilitated cogent argumentation among students. Indeed, the student teachers involved in Garrido Espeja and Couso's study tended initially to transmit specific information as established truth rather than facilitating that students could dialogically and argumentatively develop their view on the SSI. Further, they found that the participating student teachers had significant difficulties coordinating students' discussions (such difficulties seem to face in-service teachers as well; see e.g. Bryce and Gray 2004). Garrido Espeja and Couso relate these challenges to the teacher students' lack of pedagogical content knowledge as well as the lack of mastery of the scientific content. Interestingly Garrido Espeja and Couso's study also shows that the involved teacher students under careful guidance and scaffolding by teacher educators (the researchers) were able to change this practice and find more cogent ways to include scientific content.

As discussed in Chap. 2, Nielsen (2020), much research on pre-service teachers document that the development of the competence to cogently include socioscientific issues in one's teaching needs to be facilitated systematically (see also Evagorou et al. 2014b). All things being equal, becoming educated as a science teacher does not necessarily entail that one becomes competent to teach socioscientific issues.

12.3 The Embeddedness of SSI-Teaching

The second emergent theme from the chapters in this book pertains to the way SSI-teaching is often embedded in a wider teaching context with a particular pedagogical approach that often will differ from what students and/or teachers are used to. It is a well-known straw man fallacy to state that there is something called *traditional teaching* and that a particular *new* pedagogical approach differs significantly from the traditional approach. That being said, letting students discuss and make own decisions on contentious societal issues *is* different from all variations of teacher-centered teaching (we are *not* saying that the latter is the norm in all classrooms).

Coarsely put, we can expect that in many classrooms, the introduction of socioscientific issues in the science classroom will mark a palpable change in pedagogy (indeed it is well established that the introduction of socioscientific issues is rarely the norm in science teacher practices (see e.g. Lee and Witz 2009). Indeed, full-fledged SSI-teaching will inevitably involve a role in which the teacher guides students' argumentation or decision-making processes (Nielsen 2009) in which the students balance multifarious information and values coming not just from the natural sciences (Nielsen 2010).

The change in practice or difference from the (local) norm that SSI-teaching imposes is often accentuated because the introduction of socioscientific issues often occurs within the context of a wider teaching context that in itself also differs from the (local) norm. This is evident in the contributions by Amos et al. (2020), Friedrichsen et al. (2020) and Davis and Bellocchi (2020) and Mudaly (2020). Amos et al. (2020) outline an educational model to support Socioscientific Inquiry Based Learning (SSIBL) in conjunction with Citizenship Education (CE) – a model which was developed through the European project PARRISE. Here the socioscientific issues primarily serve as relevant scenarios that raise questions prompting an investigation.

12.4 The Necessity of Long-Term Professional Development

The third emergent theme from the chapters in this book pertains to the need for professional development of teachers. Several chapters in this book explicitly proceed from the vantage point that (long term) professional development of teachers is necessary in order to secure the uptake and quality of SSI teaching.

In Chap. 6, Friedrichsen et al. (2020), explore the possibilities of a collaborative professional development setup in which teachers co-design and implement SSI activities. Their work indicates that much can be gained from the process of collaborating on designing SSI-teaching, and that having multiple teachers from the same school participate holds many benefits. In Chap. 8, Cohen et al. (2020), discuss the benefits of having upper secondary school teachers participate in a professional development program that focusses on implementing inquiry-based SSI teaching (SSIBL). Their case study indicates that there is a professional progression or taxonomy, according to which teachers will first have to learn to teach SSI and then progress to learn to teach SSIBL. They further discuss the difficulty with implementing SSIBL in an educational system that does not fully formally legitimize this approach. Having available teaching materials, such as the ones developed for this program, may be a key in increasing the uptake of SSIBL teaching. In Chap. 9, Furman et al. (2020) present a long-term professional development program for in-service Argentinian teachers to support them in implementing SSI in their teaching. Their findings suggest that a long-term program can benefit from developing teachers' competences in a stepwise fashion – starting with implementing teaching activities designed by others and progressing to increasingly co-develop the activities.

The aspect that seems to emerge again and again is that SSI is rarely not included in the curricula; and that even if it is included it is not clear for the teachers how to teach and how to evaluate it (for an overview see Tidemand and Nielsen 2017).

12.5 The Future of Socioscientific Issues in Teacher Education

As the international science education community increasingly turn to the terminology of *competences* (see Ropohl et al. 2018; Rönnebeck et al. 2018), the learning goals associated with SSI teaching are obvious candidates for key competences that flesh out a Vision II (or even Vision III; see Sjöström and Eilks 2017) of scientific literacy. It seems that the community would benefit from a comprehensive overview of what skills and knowledge areas such competences are comprised of in order to establish a roadmap for potential learning goals in the realm of SSI teaching.

The use of the specific models in training, models that will include SSI competences, may help teachers to develop a stronger pedagogical base to support their teaching and learning about SSI. Until now, very few models of SSI professional development are supported by empirical data, and a contribution of this book is that we present different models from different contexts all supported by data and providing a detailed presentation of the PD context. Furthermore, the question of whether long term professional development with in class support (e.g. Bencze et al. 2020; Garrido Espeja and Couso 2020; Friedrichsen et al. 2020) has better outcomes than short term PD (Bayram-Jacobs et al. 2019) still remains. Teacher ownership and co-creation of materials (Friedrichsen et al. 2020; Garrido Espeja and Couso 2020) seems to play a positive role on how teachers uptake SSI teaching, even though some studies offer contrasting evidence (Bayram-Jacobs et al. 2019).

A notable gap in knowledge that this book has unfortunately not covered pertains to the lack of knowledge about assessment of or for student learning in SSI-teaching. Science teachers avoid assessing students' competences related to SSI – expecting that this is done in other disciplines (e.g. Steffen and Hößle 2016). They also tend to devalue SSI-relevant assessment criteria (e.g. Steffen and Hößle 2016) and they instead tend to focus on the science disciplinary content when assessing students (Christenson et al. 2017; Tidemand and Nielsen 2017). A number of authors (Ekborg et al. 2013; Evagorou et al. 2014a; Christenson et al. 2017) have started to focus on assessment of student learning in SSI-teaching, but have also generally argued that there is a significant gap in knowledge about viable assessment practices in this regard (Tidemand and Nielsen 2017). The existence of appropriate student assessment practices is a key factor in determining the uptake of concrete pedagogical approaches by teachers (Harlen 2013). This means that the uptake of SSI teaching is directly challenged by the lack of assessment practices vis à vis SSI teaching.

The chapters in this book are not different from the norm in science education research. Only one chapter of this book – Chap. 10, by Garrido Espeja and Couso (2020) – includes a focus on (formative) assessment in the design of SSI activities by pre-service teachers. But none of the chapters explore more deeply how student learning in SSI teaching can be assessed (formatively or summatively). Future research is needed to develop knowledge of how the complex learning goals associated with SSI teaching can be made *operational* (Nielsen et al. 2018) for assessment (see also Nielsen and Dolin 2016; Dolin et al. 2017).

Concluding this chapter, we believe that the book has helped in presenting current trends and successful practices in SSI teacher professional development. However, three questions still remain to be answered by the international research community: What are good ways to weave learning to teach SSI into traditional teacher education? What changes are needed to move SSI-research into providing more conclusive findings? What changes in policy and assessment are needed in order to support the uptake of full-fledged SSI-teaching?

References

Amos, R., Knippels, M.-C., & Levinson, R. (2020). Socio-scientific inquiry-based learning: Possibilities and challenges for teacher education. In *Science teacher education for responsible citizenship: Towards a pedagogy for relevance through socio-scientific issues*. Cham: Springer.

Bayram-Jacobs, D., Henze, I., Evagorou, M., Shwartz, Y., Aschim, E. L., Alcaraz-Dominguez, S., Barajas, M., & Dagan, E. (2019). Science teachers' pedagogical content knowledge development during enactment of socioscientific curriculum materials. *Journal of Research in Science Teaching, 56*(9), 1207–1233.

Bencze, L., El Hawing, S., & Zouda, M. (2020). Critical and active public engagement in addressing Socioscientific problems through science teacher education. In *Science teacher education for responsible citizenship: Towards a pedagogy for relevance through socio-scientific issues*. Cham: Springer.

Bryce, T., & Gray, D. (2004). Tough acts to follow: The challenges to science teachers presented by biotechnological progress. *International Journal of Science Education, 26*(6), 717–733.

Christenson, N., Gericke, N., & Rundgren, S. N. C. (2017). Science and language teachers' assessment of upper secondary students' socioscientific argumentation. *International Journal of Science and Mathematics Education, 15*(8), 1403–1422. https://doi.org/10.1007/s10763-016-9746-6.

Cohen, R., Zafrani, E., & Yarden, A. (2020). Science teachers as proponents of socio-scientific inquiry-based learning: From professional development to classroom enactment. In *Science teacher education for responsible citizenship: Towards a pedagogy for relevance through socio-scientific issues*. Cham: Springer.

Davis, J. P., & Bellocchi, A. (2020). Gamification of SSI's as a science pedagogy: Toward a critical rationality in teaching science. In *Science teacher education for responsible citizenship: Towards a pedagogy for relevance through socio-scientific issues*. Cham: Springer.

Dolin, J., Nielsen, J. A., & Tidemand, S. (2017). Evaluering af naturfaglige kompetencer. *Acta Didactica Norge, 11*(3), 1–28.

Ekborg, M., Ottander, C., Silfver, E., & Simon, S. (2013). Teachers' experience of working with socio-scientific issues: A large scale and in depth study. *Research in Science Education, 43*(2), 599–617. https://doi.org/10.1007/s11165-011-9279-5.

Evagorou, M., Guven, D., & Mugaloglu, E. (2014a). Preparing elementary and secondary pre-service teachers for everyday science. *Science Education International, 25*(1), 68–78.

Evagorou, M., Albe, V., Angelides, P., Couso, D., Chirlesan, G., Evans, R. H., et al. (2014b). Preparing preservice science teachers to teach socio-scientific (SSI) argumentation. *Science Teacher Education, 69*, 39–47.

Friedrichsen, P. J., Sadler, T. D., & Zangori, L. (2020). Supporting teachers in the design and enactment of socio-scientific issue based teaching in the USA. In *Science teacher education for responsible citizenship: Towards a pedagogy for relevance through socio-scientific issues*. Cham: Springer.

Furman, M., Taylor, I., Luzuriaga, M., & Podestá, M. E. (2020). Getting ready to work with socio-scientific issues in the classroom: a study with Argentine teachers. In *Science teacher education for responsible citizenship: Towards a pedagogy for relevance through socio-scientific issues*. Cham: Springer.

Garrido Espeja, A., & Couso, D. (2020). Introducing SSI in primary pre-service teacher education: scientific practices to learn the 'big ideas' of science. In *Science teacher education for responsible citizenship: Towards a pedagogy for relevance through socio-scientific issues*. Cham: Springer.

Harlen, W. (2013). *Assessment & inquiry-based science education: Issues in policy and practice*. Trieste: Global Network of Science Academies (IAP) Science Education Programme (SEP).

Lee, H., & Witz, K. G. (2009). Science teachers' inspiration for teaching socio-scientific issues: Disconnection with reform efforts. *International Journal of Science Education, 31*(7), 931–960. https://doi.org/10.1080/09500690801898903.

Leung, J. S. C., Wong, K. L., & Chan, K. K. H. (2020). Pre-service secondary science teachers' beliefs about teaching socio-scientific issues. In *Science teacher education for responsible citizenship: Towards a pedagogy for relevance through socio-scientific issues*. Cham: Springer.

Mudaly, R. (2020). Re-thinking the integration of Socioscientific issues in teacher education. In *Science teacher education for responsible citizenship: Towards a pedagogy for relevance through socio-scientific issues*. Cham: Springer.

Nielsen, J. A. (2009). Structuring students´ critical discussions through processes of decision-making on socio-scientific controversies. *Revista de Estudos Universitários, 35*(2), 139–165.

Nielsen, J. A. (2010). Functional roles of science in socio-scientific discussions. In I. Eilks & B. Ralle (Eds.), *Contemporary science education – Implications from science education research about orientations, strategies and assessment* (pp. 83–96). Aachen: Shaker.

Nielsen, J. A. (2020). Teachers and socioscientific issues – An overview of recent empirical research. In *Science teacher education for responsible citizenship: Towards a pedagogy for relevance through socio-scientific issues*. Cham: Springer.

Nielsen, J. A., & Dolin, J. (2016). Evaluering mellem mestring og præstation. [Assessment between mastery and performance]. *MONA, 2016*(1), 51–62.

Nielsen, J. A., Tidemand, S., & Dolin, J. (2018). Transforming assessment research: Recommendations for future research. In J. Dolin & R. Evans (Eds.), *Transforming assessment: Through an interplay between practice, research and policy* (pp. 279–290). Cham: Springer.

Rönnebeck, S., Nielsen, J. A., Olley, C., Ropohl, M., & Stables, K. (2018). The teaching and assessment of inquiry competences. In J. Dolin & R. Evans (Eds.), *Transforming assessment: Through an interplay between practice, research and policy* (pp. 27–52). Cham: Springer.

Ropohl, M., Nielsen, J. A., Olley, C., Rönnebeck, S., & Stables, K. (2018). The concept of competence and its relevance for science, technology and mathematics education. In J. Dolin & R. Evans (Eds.), *Transforming assessment: Through an interplay between practice, research and policy* (pp. 3–25). Cham: Springer International Publishing.

Sjöström, J., & Eilks, I. (2017). Reconsidering different visions of scientific literacy and science education based on the concept of Bildung. In Y. J. Dori, Z. Mevarech, & D. Bake (Eds.), *Cognition, metacognition, and culture in STEM education*. Dordrecht: Springer.

Steffen, B., & Hößle, C. (2016). Assessing students' performances in decision-making: Coping strategies of biology teachers. *Journal of Biological Education, 51*(1), 44–51.

Tidemand, S., & Nielsen, J. A. (2017). The role of socioscientific issues in biology teaching: From the perspective of teachers. *International Journal of Science Education, 39*(1), 44–61. https://doi.org/10.1080/09500693.2016.1264644.

Jan Alexis Nielsen is Associate Professor at the Department of Science Education at the University of Copenhagen, Denmark. His research focuses on the understanding and assessment of generic competences. He has investigated students' socioscientific argumentation and teachers' beliefs about teaching socioscientific issues. He has also worked closely with teachers to build assessment frameworks for inquiry, modelling and innovation competence. Over the past 12 years, he has participated in a range of international and national projects related to inquiry and/or socio-scientific issues.

Maria Evagorou is an Associate Professor at the University of Nicosia, Cyprus. Her scholarly activity focuses on exploring and enhancing young students' and pre-service teachers' argumentation skills within socioscientific issues. More specifically, the emphasis of her work is on students' and teachers' talk when they engage in the discussion of SSI and aims to explore ways in which teachers can be supported in their effort to include SSI in their teaching s. She has worked as a principal investigator and senior researcher on various EU and local research projects, served as a member on the JRST editorial board, and as a strand coordinator for ESERA, and has published widely.

Justin Dillon is professor of science and environmental education at the University of Exeter. After studying for a degree in chemistry, he trained as a science teacher at Chelsea College and taught in six inner London schools. Justin joined King's College London in 1989 and was promoted to professor in 2009. He is editor-in-chief of Studies in Science Education and co-edits the International Journal of Science Education. In 2007, Justin was elected President of the European Science Education Research Association. He has co-edited 18 books including Becoming a Teacher, Bad Education and the International Handbook of Research on Environmental Education. Justin has published around 100 papers in peer-reviewed journals and almost the same number of book chapters.

The manufacturer's authorised representative in the EU is Springer Nature Customer Service Centre GmbH, Europaplatz 3, 69115 Heidelberg, Germany. If you have any concerns regarding our products, please contact ProductSafety@springernature.com

Printed and bound by CPI Group (UK) Ltd, Croydon, CR0 4YY
05/10/2026
02228121-0012